T0233653

INTERNATIONAL CENTRE FOR MECHANICAL SCIENCES

COURSES AND LECTURES - No. 126

B.M. FRAEIJS de VEUBEKE
UNIVERSITY OF LIEGE

M. GERADIN
BELGIAN NATIONAL SCIENCE FOUNDATION

A. HUCK
RESEARCH ENGINEER

M.A. HOGGE
UNIVERSITY OF LIEGE

STRUCTURAL DYNAMICS

HEAT CONDUCTION

COURSES HELD AT THE DEPARTMENT
OF MECHANICS OF SOLIDS
JULY 1972

UDINE 1972

SPRINGER-VERLAG WIEN GMBH

This work is subject to copyright.

All rights are reserved,

whether the whole or part of the material is concerned

specifically those of translation, reprinting, re-use of illustrations,

broadcasting, reproduction by photocopying machine

or similar means, and storage in data banks.

© 1972 by Springer-Verlag Wien

Originally published by Springer-Verlag Wien New York in 1972

ISBN 978-3-211-81201-3 ISBN 978-3-7091-2957-9 (eBook)

DOI 10.1007/978-3-7091-2957-9

STRUCTURAL DYNAMICS

M. GERADIN

A. HUCK

B. FRAEIJS de VEUBEKE

Laboratoire de Techniques Aéronautiques

et Spatiales

Université de Liège

PREFACE

Several aspects of Structural Dynamics as they relate to the Finite Element Method are surveyed.

In Section 2 the variational principles of elastodynamics of both displacement type and stress type, on which rests the foundation of dual mathematical models of finite elements. In Section 3 the bases required for an eigenvalue analysis of a structure. In Section 4 the derivation of the classical algebraic form of eigenvalue problem using either displacement or equilibrium finite element models.

The presence of kinematical modes in the structure requires generalizations of some concepts, such as the one of flexibility matrices, which are dealt with in Section 5.

Section 6 discusses accuracy of the low frequency model analysis resulting from methods for reducing the number of degrees of freedom; it also describes bounding algorithms to assess this accuracy.

Section 7 contains numerical applications to plate like structures.

Sections 1 to 7 were written by GERADIN and FRAEYS de VEUBEKE.

In Section 8, Huck compares several methods for the numerical computations of transient response. Particular emphasis is put on the model acceleration method that exploits a low frequency model analysis.

Udine, July 1972

1. Introduction

Linear structural dynamics is one of the many field problems of engineering that can receive a variational formulation. The classical approach is the kinematical one, and the discretization of the Hamiltonian variational principle in finite elements results from a polynomial approximation of the displacement field inside each separate region. Continuity will be secured through identification of a suitable set of generalized interface displacements, in which case the kinematical elements are said to be conforming. Integrating the kinetic and potential energies of the finite elements leads to lagrangian, or so-called coherent, mass and stiffness matrices.

The first really satisfactory formulation of a dual principle, in which the kinetic energy is transformed through satisfaction of the dynamic equilibrium equations in a functional expressed in terms of an impulse field, is due to Toupin [38] . Similar approaches where followed by Crandall, Yu Chen, Gladwell and Zimmermann [6] , [41] , [24] . Some numerical applications to beam and plate problems were presented by Tabarrok, Sakaguchi and Karnopp [35] , [36] . Its use as an efficient tool in finite element applications is however very recent [19] , [21] and has still to be generalized to the three-dimensional linear elasticity.

This presentation gives a logical derivation of the dual dynamic principle through the canonical form in the spirit of the Friedrichs transformation. Its discretization results from a polynomial approximation of the stress field within each separate region. The diffusion of the boundary tractions will be preserved if a suitable system of generalized boundary loads can be chosen that may be defined uniquely in terms of the parameters of the stress field inside each element.

This paper also discusses the general procedure for assembling equilibrium, or statically admissible finite elements, in order to implement the dual principle. It will appear that, when starting from an assumed stress field, the expression of the kinetic energy of an element involves only a small number of interior parameters; the equilibrium approach leads thus in a more natural way to an "eigenvalue economizer". Zero frequency modes are associated with the other parameters, which improve the representation of the strain energy without increasing the order of the eigenvalue problems. Moreover, it is shown that there is no advantages in using the dual principle together with the requirement of forcing orthogonality with respect to all zero frequency modes. On the contrary, experience proves that, by ignoring this unnecessary requirement, the computed eigenvalues generally converge to the exact values by lower bounds, hence giving precious accuracy by comparison with the displacement approach, which always gives upper bounds.

2. THE VARIATIONAL PRINCIPLES OF ELASTODYNAMICS

2.1 Hamilton's principle

Hamilton's principle, or displacement variational principle, states that for time fixed end values of the displacement field u_i, the Lagrangian action

$$f[u] = \int_{t_1}^{t_2} (T - V)\, dt \qquad (2.1.1)$$

of a conservative system takes a stationary value on the true trajectory of the motion.

T denotes the kinetic energy of the system:

$$T = \frac{1}{2} \int_R \rho \; \dot{u}_i \, \dot{u}_i \, dR \quad , \qquad (2.1.2)$$

and V, its potential energy, can be split into distinct parts:

The strain energy

$$V_1 = \int_R W(\varepsilon)\, dR \, , \qquad (2.1.3)$$

results from the integration of the strain energy density $W(\varepsilon)$ in the domain R. If we restrict ourselves to the infinitesimal strains and rotations of a linear elastic material, W can be written as a positive definite quadratic form

$$W(\varepsilon) = \frac{1}{2} C_{ij}^{mn} \, \varepsilon_{ij} \, \varepsilon_{mn} \qquad (2.1.4)$$

of the symmetric strain tensor:

(2.1.5) $$\varepsilon_{ij} = \frac{1}{2}(D_i u_j + D_j u_i) \ .$$

The set of elastic moduli defining the tensor C^{mn}_{ij} verifies the symmetry conditions

(2.1.6) $$C^{mn}_{ij} = C^{mn}_{ji} = C^{nm}_{ij} = C^{ij}_{mn} \ .$$

The relation (2.1.5) show the strain energy to be a functional of the first order derivatives of the displacement field:

(2.1.7) $$V_1 = \int_R W(D\ u)\ dR \ .$$

According to the symmetry properties (2.1.6), we obtain the more explicit expression

(2.1.8) $$V_1 = \frac{1}{2}\int_R C^{mn}_{ij} D_i u_j D_m u_n \ dR \ .$$

Finally, we deduce from the strain energy density W the stress tensor σ_{ij} conjugated to the strain tensor ε_{ij} through the energy relation

(2.1.9) $$\sigma_{ij} = \frac{\partial W}{\partial \varepsilon_{ij}} = C^{mn}_{ij}\, \varepsilon_{mn} = C^{mn}_{ij} D_m u_n \ ,$$

under the condition that the formal distinction between ε_{ij} and ε_{ji} in the case $i \neq j$ be kept.

Another contribution to the potential energy, V_2, is a potential energy associated to the displacements on the part $\partial_2 R$ of the boundary:

$$V_2 = - \int_{\partial_2 R} F(u,t)\, d\partial R \ . \qquad (2.1.10)$$

On the remaining part $\partial_1 R$ of the boundary, we shall assume that the displacements are prescribed time functions

$$u_j = \bar{u}_j(t) \qquad \text{on} \qquad \partial_1 R \ . \qquad (2.1.11)$$

For simplicity, we do not take into consideration other potential energies like that associated with conservative external body forces functions of position (for example, gravity or electromagnetic forces).

Applying the principle

$$\delta\, F[u] = 0 \qquad (2.1.12)$$

with the Hamilton's principle rules

- at the extremities of the time interval:

$$\delta u_j = 0 \qquad \text{for} \qquad t = t_1 \qquad \text{and} \qquad t = t_2 \qquad (2.1.13)$$

- on that part of the boundary on which displacements are imposed:

$$\delta u_j = 0 \qquad \text{at all time on } \partial_1 R \ , \qquad (2.1.14)$$

the following variational derivatives are obtained

(2.1.15) $$-\rho\,\ddot{u}_j + D_i\,\sigma_{ij} = 0 \qquad \text{in } R\,.$$

They are the dynamic equilibrium conditions written in d'Alembert's form. In terms of the displacement field alone, they become

(2.1.15') $$-\rho\,\ddot{u}_j + D_i\left(C_{ij}^{mn}\,D_m u_n\right) = 0\,.$$

The natural boundary conditions supplementing the essential conditions (2.1.11) are found to be

(2.1.16) $$-n_i\,\sigma_{ij} + \frac{\partial F}{\partial u_j} = 0 \qquad \text{on } \partial_2 R\,,$$

where n_i denotes the direction cosines of the outward normal. They state that the surface tractions

(2.1.17) $$t_j = n_i\,\sigma_{ij}$$

are prescribed functions of displacement and time

(2.1.18) $$t_j = \frac{\partial F}{\partial u_j} = \bar{t}(u,t) \qquad \text{on } \partial_2 R\,.$$

An important particular case is that of a boundary potential energy which is a linear function of the displacement field:

(2.1.19) $$F(u,t) = u_j\,\bar{t}_j(t)\,.$$

The natural boundary conditions reduce then to the imposition of the surface tractions

$$t_j = \bar{t}_j(t) \qquad \text{on} \qquad \partial_2 R \ . \tag{2.1.20}$$

2.2 The canonical variational principle

We follow the way indicated by Friedrichs [2] to transform Hamilton's principle into a canonical and, later, dual form. To this purpose we introduce a dislocation potential Δ into the functional (2.1.1) that may then be written as

$$\hat{F}\left[u, v, \varepsilon, \lambda, \alpha, \mu\right] = \int_{t_1}^{t_2} (T - V + \Delta)\,dt \tag{2.2.1}$$

The dislocation potential consists of three parts: a first one

$$\Delta_1 = \int_R \lambda_{ij} \left\{ \varepsilon_{ij} - \frac{1}{2}(D_i\, u_j + D_j\, u_i) \right\} dR \ , \tag{2.2.2}$$

that removes the compatibility relations (2.1.5) as essential conditions of the variational equations for stationarity of (2.2.1) by means of a tensor of lagrangian multipliers λ_{ij} . Correspondingly, the potential energy is now expressed as

$$V_1 = \int_R W(\varepsilon)\, dR \tag{2.2.3}$$

The second part,

$$\Delta_2 = \int_{\partial_1 R} \alpha_j \left\{ u_j - \bar{u}_j(t) \right\} d\partial R \ , \tag{2.2.4}$$

incorporates the essential boundary constraints (2.1.11) as natural boundary conditions through the vector of lagrangian multipliers α_j . Finally

$$\Delta_3 = \int_R \mu_j(\dot{u}_j - v_j)\, dR \tag{2.2.5}$$

introduces the velocity field, v_j , as an independent field, and the kinetic energy is now expressed as

$$T = \frac{1}{2}\int_R \rho\, v_j v_j\, dR \,. \tag{2.2.6}$$

The introduction of the dislocation potentials Δ_1 and Δ_2 is a classical procedure used in statics to provide a logical approach to the two-field variational principles, one of which is the Reissner-Hellinger principle. Another two-field principle is given in reference [15] .

On the other hand, the introduction of the dislocation potential Δ_3 is an essential step in the logical transformation from Hamilton's principle to the eulerian variational principles of fluid mechanics [14] .

The first step in the simplification of the functional (2.2.1) consists in identifying the multipliers. This results immediately from setting the variational derivatives of ε_{ij} and v_j equal to zero:

$$\lambda_{ij} - \frac{\partial W}{\partial \varepsilon_{ij}} = 0 \tag{2.2.7}$$

and

$$\mu_i - \rho\, v_i = 0 \qquad \text{in} \quad R\,, \tag{2.2.8}$$

and

$$\alpha_j - n_i\, \sigma_{ij} = 0 \qquad \text{on} \quad \partial_1 R\,. \tag{2.2.9}$$

Hence the λ_{ij} tensor is identified with the stress tensor σ_{ij} related to the strain tensor by the constitutive equation (2.1.9); similarly, α_j is identified through (2.1.18) with the surface tractions t_j on the part $\partial_1 R$ of the boundary.

On the other hand, (2.2.8) shows that the μ_i vector corresponds to the momentum per unit mass.

When (2.2.7) is substituted into the functional (2.2.1), the expression

$$\varepsilon_{ij} \frac{\partial W}{\partial \varepsilon_{ij}} - W \tag{2.2.10}$$

that appears in the intergrand of the functional is treated as a Legendre transformation introducing the complementary energy density $\Phi(\sigma)$, a function of the elements of the stress tensor. This transformation is possible whenever the strains can be expressed in terms of the stresses, in which case, through differentiation of

$$\varepsilon_{ij}\, \sigma_{ij} - W = \Phi(\sigma)\,, \tag{2.2.11}$$

the dual constitutive equations

$$\varepsilon_{ij} = \frac{\partial \Phi}{\partial \sigma_{ij}} \tag{2.2.12}$$

follow immediately.

We thus obtain

$$\Delta_1 - V_1 = U_1 - \Gamma_1 \tag{2.2.13}$$

with the complementary strain energy

$$U_1 = \int_R \Phi(\sigma)\, dR \quad , \tag{2.2.14}$$

and

$$\Gamma_1 = \int_R \sigma_{ij} D_i u_j \, dR \tag{2.2.15}$$

where account was taken of the symmetry of the stress tensor. Also the substitution of (2.2.8) yields

$$T + \Delta_3 = -T + \Gamma_3 \tag{2.2.16}$$

where

$$\Gamma_3 = \int_R \rho\, v_j \dot{u}_j \, dR \ . \tag{2.2.17}$$

The meaning (2.2.9) of the multipliers α_j on the part $\partial_1 R$ of the boundary is utilized in order to transform Δ_2 into

$$\Delta_2 = \int_{\partial_1 R} n_i \sigma_{ij} \left\{ u_j - \bar{u}_j(t) \right\} d\partial R \ . \tag{2.2.18}$$

The final result is a canonical variational principle in the Friedrichs sense, requiring the stationarity of the functional

$$c\left[u, v, \sigma\right] = \int_{t_1}^{t_2} (U_1 + T - V_2 + \Delta_2 - \Gamma_1 + \Gamma_3)\, dt \qquad (2.2.19)$$

in which the various integrands are given respectively by (2.2.14), (2.2.6), (2.1.10), (2.2.18), (2.2.15) and (2.2.17). It can be used to develop approximate solutions based on independent discretizations of displacements, velocities and stresses. Taking the variational derivatives of (2.2.19) restitutes the natural conditions of the canonical principle. The Euler equations are respectively:

-the dynamic equilibrium equations

$$D_i \sigma_{ij} - \rho \dot{v}_j = 0 \qquad \text{in} \quad R \qquad (2.2.20)$$

and

$$t_j = \bar{t}_j(u, t) \qquad \text{on} \quad \partial_2 R \qquad (2.2.21)$$

for the variations on displacements;

-the compatibility equations (2.1.5) written in terms of the stress tensor

$$\frac{\partial \Phi}{\partial \sigma_{ij}} - \frac{1}{2}(D_i u_j + D_j u_i) = 0 \qquad \text{in } R \qquad (2.2.22)$$

and

(2.2.23) $$u_i = \bar{u}_i(t) \qquad \text{on } \partial_1 R$$

for the variations on stresses;

-the constraints between velocity and displacement fields

(2.2.24) $$\rho(\dot{u} - v_i) = 0$$

for the variations on velocities.

Note that all boundary conditions are now natural. For finite element applications of the variational principles, it is important to recognize the nature of the transition conditions at interfaces. Because the only space derivatives contained in the functional are those affecting displacements, we must consider that physical continuity

(2.2.25) $$(u_j)_+ = (u_j)_-$$

is an a priori requirement; the natural transition conditions provided by the principle and deduced from the common variations at interfaces are then

(2.2.26) $$(t_j)_+ + (t_j)_- = 0 \ .$$

From that point of view, the canonical principle behaves as the classical principle of variation of displacements. Obviously the interface constraints could be incorporated into the principle through an additional dislocation potential and the corresponding multipliers identified. Then, just as was found to be the case for the boundary conditions, all transitional conditions would be cared for by the principle.

2.3 The complementary energy principle of elastodynamics

The complementary energy principle of elastodynamics can easily be deduced from the canonical principle by specializations.

To this purpose, the functional (2.2.19) is integrated by parts with respect to the displacement field in both the terms Γ_1 and Γ_3 of the integrand.

The first transformation

$$\Gamma_1 = \int_{\partial R} n_i \sigma_{ij} u_j \, d\partial R - \int_R u_j D_i \sigma_{ij} \, dR \qquad (2.3.1)$$

cancels a part of the contribution of Δ_2 :

$$- V_2 + \Delta_2 - \Gamma_1 = \int_{\partial_2 R} \left\{ F(u,t) - n_i \sigma_{ij} u_j \right\} d\partial R$$
$$- \int_{\partial_1 R} n_i \sigma_{ij} \bar{u}_j \, \partial dR + \int_R u_j D_i \sigma_{ij} \, dR \ . \qquad (2.3.2)$$

In the second transformation,

$$\int_{t_1}^{t_2} \Gamma_3 \, dt = \left[\int_R \rho \, v_j u_j \, dR \right]_{t_1}^{t_2} - \int_{t_1}^{t_2} \int_R \rho \, \dot{v}_j u_j \, dR \, dt \qquad (2.3.3)$$

the term at the time limits appearing in the right-hand side is dropped on the basis of the Hamilton's rule (2.1.13), and the introduction of the new requirements

(2.3.4) $$\delta v_j = 0 \quad \text{for} \quad t = t_1 \quad \text{and} \quad t = t_2 .$$

The functional has thus been modified as follows:

$$c[u, v, \sigma] = \int_{t_1}^{t_2} \Big[\int_R \big\{ (D_i \sigma_{ij} - \rho \dot{v}_j) u_j \big\} dR + U_1 - T$$
$$+ \int_{\partial_2 R} \big\{ F(u,t) - t_j u_j \big\} d\partial R - \int_{\partial_1 R} t_j \bar{u}_j \, d\partial R \Big] dt ,$$

(2.3.5)

and may be simplified by making the assumption that the dynamic equilibrium equations (2.2.20) are a priori satisfied; this causes the first term to vanish. On the other hand, the integrand of the term $\partial_2 R$ is obviously related to another Legendre transformation

(2.3.6) $$u_j \frac{\partial F}{\partial u_j} - F(u,t) = G(t_j, t)$$

which is defined whenever the surface tractions given by (2.1.19) depend on the displacement field, in which case the Hessian matrix

$$\left[\frac{\partial^2 F}{\partial u_i \, \partial u_j} \right] \qquad (2.3.7)$$

does not vanish. The complementary function is then such that

$$\frac{\partial G}{\partial t_j} = u_j \quad . \qquad (2.3.8)$$

With the notations

$$B_1 = \int_{\partial_1 R} t_j \, \bar{u}_j(t) \, d\partial R \qquad (2.3.9)$$

and

$$B_2 = \int_{\partial_2 R} G(t_j, t) \, d\partial R \quad , \qquad (2.3.10)$$

the canonical functional reduces now to one that depends only on the stress field:

$$g[\sigma] = \int_{t_1}^{t_2} (U_1 - T - B_1 - B_2) \, dt \qquad (2.3.11)$$

provided the kinetic energy be also expressed in terms of the stress field. The dynamical equilibrium equations (2.2.20) that must be satisfied a priori allow precisely to express the velocity field in terms of stresses:

$$v_j = v_j \big|_{t_1} + \frac{1}{\rho} D_i \int_{t_1}^{t} \sigma_{ij} \, dt \qquad (2.3.12)$$

The time integration is avoided by introducing an impulse field, τ_{ij} , such that

(2.3.13) $$v_j = \frac{1}{\rho} D_i \tau_{ij} .$$

Hence we adopt the definition

(2.3.14) $$\tau_{ij} = \tau_{ij}\Big|_{t_1} + \int_{t_1}^{t} \sigma_{ij} \, dt$$

which yields

(2.3.15) $$\sigma_{ij} = \dot{\tau}_{ij} \quad ,$$

and we have to replace the requirements on the velocity field at time limits by

(2.3.16) $$\delta \tau_{ij} = 0 \qquad t = t_1 \qquad t = t_2$$

A convenient formulation of the complementary energy principle of elastodynamics is thus the stationarity of the functional

$$g[\tau] = \int_{t_1}^{t_2} \Big[\int_R \Big\{ \Phi(\dot{\tau}_{ij}) - \frac{1}{2\rho} D_i \tau_{ij} D_m \tau_{mj} \Big\} dR$$

(2.3.17) $$- \int_{\partial_1 R} t_j \bar{u}_j(t) \, d\partial R - \int_{\partial_2 R} g(t_j, t) \, d\partial R \Big] dt .$$

It depends only on the impulse field τ_{ij} , since the associated surface tractions can be written as

(2.3.18) $$t_j = n_i \dot{\tau}_{ij} \quad .$$

The variational derivatives of this principle are

$$-\frac{d}{dt}\frac{\partial \Phi}{\partial \dot{\tau}_{ij}} + \frac{1}{2\rho}(D_i D_m \tau_{mj} + D_j D_m \tau_{mi}) = 0 \qquad (2.3.19)$$

that, in view of (2.3.15), (2.2.12) and (2.2.20), are easily interpreted as the time derivatives of the compatibility conditions (2.1.5):

$$\dot{\varepsilon}_{ij} - \frac{1}{2}(D_i v_j + D_j v_i) = 0 \ . \qquad (2.3.20)$$

The natural boundary conditions are seen to be

$$-\frac{1}{\rho} D_i \tau_{ij} + \frac{d}{dt}\bar{u}_j(t) = 0 \quad \text{on} \quad \partial_1 R \qquad (2.3.21)$$

$$-\frac{1}{\rho} D_i \tau_{ij} + \frac{d}{dt}\frac{\partial G}{\partial \dot{\tau}_{ij}} = 0 \quad \text{on} \quad \partial_2 R \ . \qquad (2.3.22)$$

Both are easily interpreted through (2.3.13) and (2.3.8) as expressing the continuity of the velocity field on boundaries. In the special case where the imposed surface tractions (2.1.19) do not depend on the displacement field, the Legendre transformation (2.3.6) yields a G function that is identically zero; the boundary requirement

$$t_j = \bar{t}_j(t) \qquad \text{on } \partial_2 R \qquad (2.3.23)$$

becomes then an essential condition in the complementary energy formulation. Similarly, because the displacement field vanishes completely, the continuity of the impulse field (2.2.26) becomes an a priori requirement.

2.4 Matrix formulation of the variational principles

Displacements, surface tractions, strains, stresses and impulses can be presented as row vectors:

$$u' = (u_1\ u_2\ u_3)$$

$$t' = (t_1\ t_2\ t_3)$$

(2.4.1)
$$\varepsilon' = (\varepsilon_{11}\ \ \varepsilon_{22}\ \ \varepsilon_{33}\ \ \gamma_{23}\ \ \gamma_{31}\ \ \gamma_{12})$$

$$\sigma' = (\sigma_{11}\ \ \sigma_{22}\ \ \sigma_{33}\ \ \sigma_{23}\ \ \sigma_{31}\ \ \sigma_{12})$$

$$\tau' = (\tau_{11}\ \ \tau_{22}\ \ \tau_{33}\ \ \tau_{23}\ \ \tau_{31}\ \ \tau_{12})$$

with the classical definition of the shearing strains:

(2.4.2)
$$\gamma_{ij} = \gamma_{ji} = \varepsilon_{ij} + \varepsilon_{ji} \ .$$

Then, with the help of a matrix differential operator

(2.4.3)
$$D' = \begin{pmatrix} D_1 & 0 & 0 & 0 & D_3 & D_2 \\ 0 & D_2 & 0 & D_3 & 0 & D_1 \\ 0 & 0 & D_3 & D_2 & D_1 & 0 \end{pmatrix}$$

and a corresponding matrix of direction cosines for the outward normal

$$N' = \begin{pmatrix} n_1 & 0 & 0 & 0 & n_3 & n_2 \\ 0 & n_2 & 0 & n_3 & 0 & n_1 \\ 0 & 0 & n_3 & n_2 & n_1 & 0 \end{pmatrix}, \qquad (2.4.4)$$

The basic equations of linear elasticity theory recalled in the preceding sections take the following forms

$\varepsilon = D\,u$ strain-displacement relation (2.4.5)

$D'\,\tau - \rho\,\ddot{u} = 0$ dynamic equilibrium equations (2.4.6)

$t = N'\sigma$ surface equilibrium equations (2.4.7)

The linear stress-strain relations

$$\sigma = H\varepsilon \qquad\qquad \varepsilon = H^{-1}\sigma \qquad (2.4.8)$$

involve a symmetric, positive - definite matrix H of elastic moduli.

Following these definitions, the functionals of both displacement and complementary energy principles may be written in matrix form:

$$f[u] = \int_{t_1}^{t_2} \left[\frac{1}{2} \int_R \rho\, \dot{u}' \dot{u}\, dR - \frac{1}{2} \int_R (Du)' H (Du)\, dR + \int_{\partial_2 R} \bar{t}'\, u\, d\partial R \right] dt, \tag{2.4.9}$$

and

$$g[\tau] = \int_{t_1}^{t_2} \left[\frac{1}{2} \int_R \dot{\tau}' H^{-1} \dot{\tau}\, dR - \frac{1}{2} \int_R \frac{1}{\rho} (D'\tau)'(D'\tau)\, dR - \int_{\partial_1 R} (N'\dot{\tau})'\, \bar{u}\, d\partial R \right] dt \tag{2.4.10}$$

in the particular case where the external potential energy results from the imposition of the surface tractions on $\partial_1 R$.

The matrix notation adopted here is more suitable than tensor notations when discretizing the functionals (2.4.9) and (2.4.10) by the finite element method.

3. EIGENVALUE ANALYSIS

The computation of the eigenvalues and associated eigenmodes of an elastic structure can be considered as one of the most important goals aimed at in the dynamic analysis of structures.

The benefit expected from a dual approach of the eigenvalue problem is the same as that guaranteed in statics:

a bracketing of the exact solution between upper and lower bounds when using alternately the kinematical and equilibrium approaches. Keeping the eigenvalue problem under variational form is the easiest way to derive the respective properties of both analyses.

3.1 The kinematical approach

3.3.1 The Rayleigh quotient

Under the assumption that no external load is applied to the system, the Euler equation of the displacement variational principle reduces to

$$D'(H\,Du) - \rho\,\ddot{u} = 0 \tag{3.1.1}$$

under the a priori requirement

$$u = 0 \qquad \text{on} \qquad \partial_1 R \tag{3.1.2}$$

that defines the set of admissible solutions. The natural boundary conditions obtained from the principle are homogeneous

$$t = N'\sigma = 0 \qquad \text{on} \qquad \partial_2 R \tag{3.1.3}$$

As is well known, the general solution can be obtained by superposition of harmonic solutions of the form

(3.1.4) $$u(x,t) = u(x)\cos(\omega t + \Phi)$$

obtained by separation of the time variable.

The eigen-modes and eigenvalues of the corresponding self-adjoint differential system

(3.1.5) $$\left.\begin{array}{l} D'(H\,D u) + \omega^2 \rho\, u = 0 \\ u = 0 \quad \text{on} \quad \partial_1 R \end{array}\right\}$$

are denoted by

(3.1.6) $$\left\{\begin{array}{l} u_{(1)},\, u_{(2)},\, \dots,\, u_{(n)} \dots \\ \omega_1^2 \leqslant \omega_2^2 \leqslant \dots \leqslant \omega_n^2 \dots \end{array}\right.$$

We can define the corresponding Rayleigh quotient

(3.1.7) $$\left\{\begin{array}{l} \omega^2 = R(u) = \dfrac{\int_R (D u)' H (D u)\, dR}{\int_R \rho\, u' u\, dR} \\ u = 0 \qquad \text{on} \qquad \partial_1 R \end{array}\right.$$

and by taking variations with respect to the displacement vector u , it follows immediately that (3.1.7) takes a stationary value $\omega^2 = \omega_i^2$ for the corresponding eigenmode $u_{(i)}$.

We recall the well known orthogonality properties:

$$\int_R (D u_{(i)})' H (D u_{(j)}) dR = \gamma_i \, \delta_{ij} \tag{3.1.8}$$

and

$$\int_R \rho \, u'_{(i)} u_{(j)} \, dR = \mu_i \, \delta_{ij} \, . \tag{3.1.9}$$

The generalized stiffness and mass so introduced, γ_i and μ_i, are not independent of each other, since (3.1.7) gives

$$\omega_i^2 = \frac{\gamma_i}{\mu_i} \, . \tag{3.1.10}$$

A convenient choice of norm consists thus to impose

$$\mu_i = \int_R \rho \, u'_{(i)} u_{(i)} \, dR = 1 \tag{3.1.11}$$

so that

$$\int_R (D u_{(i)})' H (D u_{(i)}) \, dR = \omega_i^2 \, . \tag{3.1.12}$$

Note that the possible zero frequency modes of the structure have to be interpreted as either the rigid body modes or the mechanism of the structure, since (3.1.12) shows that they involve no strain energy.

3.1.2 Independent, or maximum-minimum characterization of eigenvalues [2]

All minimum properties of the Rayleigh quotient in the displacement approach can be deduced from Courant's theorem, often called the "minimax principle".

To this purpose let us turn to the variational problem of minimizing the Rayleigh quotient (3.1.7) under arbitrary constraints taking the form of conditions of orthogonality:

$$(3.1.13) \qquad \int_R v'_j \, u \, dR = 0 \qquad\qquad j = 1, \dots, r-1$$

Courant's principle states that "the rth eigenvalue of problem (3.1.5) is the maximum value which can be given to the minimum of the Rayleigh quotient (3.1.7) by varying the $r-1$ constraints (3.1.13). This maximum is reached for $v_j = \rho \, u_{(j)}$".

If we denote by $m(v_1, \dots v_{r-1})$ the constrained minima of the Rayleigh quotient, Courant's principle can thus be written as

$$(3.1.14) \qquad \omega_r^2 = \max m(v_1, \dots v_{r-1}) = m(\rho \, u_{(1)}, \dots, \rho \, u_{(r-1)})$$

To prove it, express the admissible solutions as expansions in eigenmodes of the unconstrained system:

$$(3.1.15) \qquad u = \sum_{i=1}^{\infty} a_i \, u_{(i)} \ .$$

Thanks to the orthonorming properties (3.1.8) to (3.1.12), the problem is transformed into

$$(3.1.16) \qquad \omega^2 = \frac{\sum_1^{\infty} a_i^2 \, \omega_i^2}{\sum_1 a_i^2} \qquad \text{minimum}$$

under the constraints

$$\sum_{i=1}^{\infty} c_{ij} a_i = 0 \qquad j = 1, 2 \ldots r-1 \qquad (3.1.17)$$

where

$$c_{ij} = \int_R v'_j \, u_{(i)} \, dR \; . \qquad (3.1.18)$$

Or, equivalently,

$$m = \sum_1^{\infty} a_i^2 \omega_i^2 \qquad \text{minimum} \qquad (3.1.19)$$

under the constraints (3.1.17) and the additional norming constraint

$$\sum_1^{\infty} a_i^2 = 1 \qquad (3.1.20)$$

Under given constraints (3.1.17) m takes the value

$$m = \sum_1^{r} a_i^2 \omega_i^2 \leqslant \omega_r^2 \sum_1^{r} a_i^2 = \omega_r^2 \qquad (3.1.21)$$

for the particular set of values

$$a_i = 0 \qquad i > r$$

and values of a_i for $i < r$ satisfiying the constrains, which are then reduced to

$$(3.1.22)\quad \sum_{1}^{r} c_{ij} a_i = 0 \qquad j = 1,2\ldots r-1$$

$$(3.1.23)\quad \sum_{1}^{r} a_i^2 = 1$$

The homogeneous system (3.1.22) has always a non trivial solution (the number of unknowns exceeds by one the number of equations), that can be scaled to satisfy (3.1.23). The result (3.1.21) being independent of the constraints, it can be concluded that, whatever those may be, the constrained minimum cannot exceed the rth eigenvalue ω_r^2 . Courant's principle is proved if a set of constraints can be produced for which the minimum actually reaches ω_r^2 ; it is then the maximum of all minimums. This is the case for the choice

$$v_j = \rho\, u_{(j)} \qquad j = 1,2\ldots r-1$$

for then we find from the orthonorming properties

$$c_{ij} = \delta_{ij} \text{ and, in particular } c_{ij} = 0 \text{ for } i \geqslant r.$$

Thus the constraints (3.1.17) require simply that

$$a_i = 0 \qquad i = 1,2\ldots r-1$$

The choice

$$a_r = 1 \qquad a_i = 0 \qquad i > r.$$

satisfies the norming condition (3.1.20) and we find

$$m = \omega_r^2 .$$

3.1.3 Recursive characterization of eigenvalues

As a direct consequence of the minimax principle, we consider the restricted class of admissible solutions which consists only of displacement modes orthogonal to the first $r-1$ eigenvectors

$$\int_R \rho\, u_{(j)}' u \, dR = 0 \qquad j = 1, \ldots r-1. \qquad (3.1.24)$$

Then

$$\min_u R(u) = \omega_r^2 , \qquad (3.1.25)$$

and this minimum is reached for $u = u_{(r)}$. This well known result defines an efficient procedure for obtaining eigenvalues recursively by minimization techniques [18] .

3.1.4 Minimum of the Rayleigh quotient under constraints

The maximum minimum property of eigenvalues also

allows to predict how the eigenspectrum will be modified by imposing a set of independent constraints:

(3.1.26) $$\int_R g'_j \, u \, dR = 0 \qquad j = 1, \dots s$$

As a first consequence of Courant's principle, ω_r^2 , is the maximum value which can be given to the minimum of $R(u)$ when varying the $r-1$ arbitrary constraints (3.1.13):

(3.1.27) $$\omega_r^2 = \max_{v} m(v_1, \dots v_{r-1}).$$

Next denote by $\tilde{\omega}_r^2$ the rth eigenvalue of the constrained Rayleigh quotient. We may also write

(3.1.28) $$\tilde{\omega}_r^2 = \max_{v} m(v_1 \dots v_{r-1} ; g_1, \dots g_s),$$

which is obviously bounded from below by the corresponding eigenvalue of the unrestrained minimum problem.

Hence

(3.1.29) $$\omega_r^2 \leqslant \tilde{\omega}_r^2$$

Now let us also vary the s imposed constraints (3.1.26): the maximum value that would be reached by the minimum of $R(u)$ if all $r+s-1$ constraints could be varied is equal to

(3.1.30) $$\omega_{r+s}^2 = \max_{v,g} m(v_1, \dots v_{r-1} ; g , \dots g_s)$$

Hence we obtain the second inequality

$$\tilde{\omega}_r^2 \leqslant \omega_{r+s}^2 \tag{3.1.31}$$

We have thus established Rayleigh's theorem on constraints:

"If s arbitrary constraints are imposed on a vibrating system of which eigenvalues are given by (3.1.6), then the new eigenvalues $\tilde{\omega}_r^2$ separate the old ones in the sense that

$$\omega_r^2 \leqslant \omega_r^2 \leqslant \omega_{r+s}^2 \tag{3.1.32}$$

for every $r \leqslant n - s$

Note that the discretization of the eigenvalue differential system (3.1.5), as it reduces the class of admissible functions to a finite set of displacement modes, may be considered as an imposition of additional constraints to the eigenvalue problem. Rayleigh's theorem guarantees thus the upper bound character of eigenfrequencies computed from a kinematical approach.

3.2 The equilibrium approach

3.2.1 Self-stressing and vibration modes

The variational derivative of the complementary principle (2.4.10) is

(3.2.1) $$-H^{-1}\ddot{\tau} + D\left(\frac{1}{\rho}D'\tau\right) = 0$$

and its natural boundary condition

(3.2.2) $$-\frac{1}{\rho}D'\tau + \frac{d}{dt}\bar{u} = 0 \qquad \text{on } \partial_1 R$$

to which we must add the a priori boundary condition

(3.2.3) $$t = N'\dot{\tau} = 0 \qquad \text{on } \partial_2 R$$

In the spectral analysis problem we set

(3.2.4) $$\tau(x,t) = \tau(x)\cos(\omega t + \Phi)$$

whereby the Euler equation (3.2.1) takes the form

(3.2.5) $$\omega^2 H^{-1}\tau + D(\frac{1}{\rho}D'\tau) = 0$$

The Rayleigh-type quotient appears in the form

(3.2.6) $$\omega^2 = R(\tau) = \frac{\int_R \frac{1}{\rho}(D'\tau)'D'\tau\, dR}{\int_R \tau' H^{-1}\tau\, dR}$$

in which the amplitude vector $\tau(x)$ must satisfy the a priori boundary condition

(3.2.7) $$N'\tau = 0 \qquad \text{on } \partial_2 R$$

As in the kinematical approach $\omega^2 = 0$ belongs to the spectrum of eigenvalue problem (3.2.5) and (3.2.6). It appears clearly on (3.2.6) that the eigenmodes pertaining to this eigenvalues satisfy

$$D'\tau = 0 \tag{3.2.8}$$

together with (3.2.7). They are statical self-stressing modes, that is stress distributions in equilibrium without external forces; thus capable of existing without inertia forces. In contrast however to the kinematical approach, their number is infinite for a continuum (finite but large after a discretization), while the kinematical modes are finite in number and even non existent if the structure is at least isostatically supported.

The self-stressing modes will be denoted by $\sigma_{(i)}$ and will be supposed to be referred to an orthonormal basis

$$\int_R \sigma'_{(i)} H^{-1} \sigma_{(j)} \, dR = \delta_{ij} \quad . \tag{3.2.9}$$

The other eigenmodes will be denoted by $\tau_{(r)}$ and ranged in increasing order of their eigenvalues

$$\left.\begin{array}{l} \tau_{(1)} \;,\; \tau_{(2)} \;, \ldots \quad \tau_{(n)} \;\ldots \\ \omega_1^2 \;\leq\; \omega_2^2 \quad \ldots \leq \omega_n \;\ldots \end{array}\right\} . \tag{3.2.10}$$

We may assume without loss of generality that, in addition to (3.2.9), we dispose of the orthonorming properties

(3.2.11) $$\int_R \sigma'_{(i)} H^{-1} \tau_{(r)} \, dR = 0$$

(3.2.12) $$\int_R \tau'_{(r)} H^{-1} \tau_{(s)} \, dR = \delta_{rs}$$

(3.2.13) $$\int_R (D' \tau_{(r)})' D' \tau_{(s)} \, dR = \omega_r^2 \delta_{rs}$$

3.2.2 Properties of the Rayleigh quotient [13] , [20]

As a consequence of Courant's principle, the Rayleigh quotient will theoretically furnish an upper bound to the eigenvalue ω_1^2 , provided τ is orthogonal to all self-stressing modes of zero frequency.

It is of interest to note that it is quite possible to construct impulse distributions that are orthogonal to all the self-stressing modes. Orthogonality means

$$\int_R \sigma'_{(i)} \hat{\varepsilon} \, dR = 0$$

where $\sigma_{(i)}$ is any self-stressing mode, and $\hat{\varepsilon}$ is the strain distribution associated with the orthogonal stress distribution through the constitutive equations. A sufficient condition to implement (3.2.14) is to make the strain distribution compatible:

$$\hat{\varepsilon} = D\,\hat{u} \qquad \text{in} \quad R \tag{3.2.15}$$

with the homogeneous boundary condition

$$\hat{u} = 0 \qquad \text{on} \quad \partial_1 R\,. \tag{3.2.16}$$

Indeed one obtains then from integration by parts

$$\int_R \sigma'_{(i)}\,\hat{\varepsilon}\,dR = \int_R \sigma'_{(i)}(D\,\hat{u})\,dR$$

$$= \int_{\partial R} (N'\sigma_{(i)})'\hat{u}\,dR - \int_R (D'\sigma_{(i)})'\hat{u}\,dR\,,$$

and the right-hand side vanishes in view of (3.2.7) and (3.2.8), which are applicable to $\sigma_{(i)}$, and in view of (3.2.16). The condition is also necessary as can be shown by satisfying (3.2.8) with the help of stress functions, integrating (3.2.14) by parts and using the arbitrariness in the stress functions to show that the strains have to satisfy the integrability conditions for the existence of displacements.

In view of

$$\tau = H\,\hat{\varepsilon} = H(D\,\hat{u})\,, \tag{3.2.17}$$

the Rayleigh quotient becomes expressible entirely in terms of the auxiliary displacement field $\hat{u}$:

$$R(\hat{u}) = \frac{\int_R \frac{1}{2}(D'H\,D\,\hat{u})'(D'H\,D\,\hat{u})\,dR}{\int_R (D\,\hat{u})'(D\,\hat{u})\,dR} \tag{3.1.18}$$

This new Rayleigh quotient, that leads immediately to the non zero eigenvalues of the problem, has the same kinematical boundary conditions (3.2.16) to be satisfied ab initio as (3.1.7), but in addition the transformed boundary conditions (3.2.7)

$$N'H(D\,\hat{u}) = 0 \qquad \text{on} \quad \partial_2 R\,. \tag{3.1.19}$$

An interesting inequality [13] relates (3.1.7) to (3.2.18). To prepare its proof consider the following obvious inequality

$$\int_R \left(\sqrt{\rho}\,a - \frac{\lambda b}{\sqrt{\rho}}\right)'\left(\sqrt{\rho}\,a - \frac{\lambda b}{\sqrt{\rho}}\right) dR \geq 0 \tag{3.1.20}$$

valid for any value of the scalar λ . Expanding it gives

$$\int_R \rho\, a'a\,dR - 2\lambda \int_R a'b\,dR + \lambda^2 \int_R \frac{1}{\rho}\, b'b\,dR \geq 0\,. \tag{3.2.21}$$

and the minimum of the left-hand side is reached for

$$\lambda = \frac{\int_R a'\,b\,dR}{\int_R \frac{1}{\rho}\, b'\,b\,dR}\,. \tag{3.2.22}$$

Hence, after substitution

$$\int_R \frac{1}{2}\rho\, a'a\, dR - \left\{\int_R \frac{1}{2} a'b\, dR\right\}^2 \left\{\int_R \frac{1}{2\rho} b'b\, dR\right\}^{-1} \geq 0 \quad (3.2.23)$$

setting

$$a = \hat{u} \quad (3.2.24)$$

and

$$b = D'(H \quad D\hat{u}) \quad (3.2.25)$$

we already identify two of the integrals as the denominator of (3.1.7) and the numerator of (3.2.18).

For the third integral

$$\int_R \frac{1}{2} a'b\, dR = \int_R \frac{1}{2} \hat{u}'D'(H \quad D\hat{u})\, dR$$

$$= \int_{\partial R} \frac{1}{2} (N\hat{u})'H(D\hat{u})\, dS - \int_R \frac{1}{2} (D\hat{u})'H(D\hat{u})\, dR$$

$$= - \int_R \frac{1}{2} (D\hat{u})'H(D\hat{u})\, dR \quad (3.2.26)$$

since the surface integral vanishes on account of (3.2.16) and (3.2.19). We obtain thus the inequality

$$\omega_R^2 \leq \hat{\omega}_R^2$$

where ω_R^2 denotes the classical Rayleigh quotient (3.1.7), and $\hat{\omega}_R^2$, the quotient (3.2.8). This means that whenever a displacement field is chosen, that satisfies both the kinematical and the stress boundary conditions, the classical Rayleigh quotient is always a better approximation than the one derived from the stress approach with orthogonality to all zero frequency modes. However, as numerical experience shows, direct applications of (3.2.18) usually under-estimate the frequency and converge through lower bounds when the number of degrees of freedom is increased. This is of course due to the fact that the assumed stress modes do not satisfy orthogonality with respect to all the self-stressing modes. Unfortunately there is no guarantee of this property of lower boundness, and further theoretical research is necessary to be able to incorporate this into the formalism.

4. FINITE ELEMENT MODELS

4.1 The displacement approach [12], [21]

Element stiffness and matrices

Consider the simply connected domain E of an element bounded by its surface ∂E on which it will be convenient to assume, for the time being, that all surface tractions are specified.

The displacement field within the element will

in general be discretized in terms of polynomials contained in a $3 \times n(a)$ matrix $P(x)$

$$u(x,t) = P(x)\, a(t) \tag{4.1.1}$$

where $a(t)$ is a column matrix of unknown time dependent coefficients. It is important for purposes of connections between elements to distinguish between the displacement field along the boundary of the element and in its interior. The boundary displacement field will be determined unambiguously by a set of generalized boundary coordinates. When those are chosen, we always obtain their values in terms of the set of coefficients

$$q(t) = M\, a(t) \tag{4.1.2}$$

We shall assume, temporarily, that the homogeneous adjoint equation

$$M'g = 0$$

has but the trivial solution zero. This ensures that the q can be chosen independently and that the general inverse of (4.1.2) is of type

$$a(t) = Q\, q(t) + B\, b(t) \tag{4.1.3}$$

where the first term is a particular solution, the second is the general solution of $Ma = 0$ and contains the arbitrary column matrix b . Substitution of (4.1.3) into (4.1.1) yields

$$u(x,t) = Q(x)\,q(t) + B(x)\,b(t) \tag{4.1.4}$$

with

$$Q(x) = P(x)\,Q \qquad\qquad B(x) = P(x)\,B \tag{4.1.5}$$

The $Q(x)$ are the "shaping functions", the $B(x)$ the "bubble functions" so-called because they vanish at the boundary, since the boundary displacements are, by assumption, uniquely determined by the q coordinates. When there are bubble functions, the shaping functions are not unique, they can be modified by arbitrary additions of bubble functions; this is however not essential for the developements to be presented here. The strain field can be represented indifferently by

$$\begin{aligned} \varepsilon(x,t) &= D\,u(x,t) = \left(D\,P(x)\right)\,a(t) \\ &= \left(D\,Q(x)\right)\,q(t) + \left(D\,B(x)\right)\,b(t) \end{aligned} \tag{4.1.6}$$

Discretization of the strain energy is similarly

$$V_1 = \int_E W(\varepsilon)\,dE = \frac{1}{2}\int_E \varepsilon' H\,\varepsilon\,dE = \frac{1}{2}\,a' K_{aa}\,a \tag{4.1.7}$$

$$= \frac{1}{2}\,q' K_{qq}\,q + q' K_{qb}\,b + \frac{1}{2}\,b' K_{bb}\,b \tag{4.1.8}$$

In general

$$K_{aa} = \int_E \left(D\,P(x)\right)' H \left(D\,P(x)\right) dE \tag{4.1.9}$$

is easier to compute and

$$\left.\begin{aligned} K_{qq} &= \int_E \left(DQ(x)\right)' H \left(DQ(x)\right) dE \\ K_{qb} &= \int_E \left(DQ(x)\right)' H \left(DB(x)\right) dE \\ K_{bb} &= \int_E \left(DB(x)\right)' H \left(DB(x)\right) dE \end{aligned}\right\} \tag{4.1.10}$$

instead of being calculated from the derivatives of shaping and bubble functions can also be obtained from (4.1.3) and K_{aa}

$$V_1 = \frac{1}{2}(b'B' + q'Q')\,K_{aa}(Qq + B\,b)$$

$$K_{qq} = Q'K_{aa}Q \qquad K_{qb} = Q'K_{aa}B = K'_{bq} \tag{4.1.11}$$

$$K_{bb} = B'K_{aa}\,B$$

Similarly the kinetic energy discretization

(4.1.12) $$T = \frac{1}{2}\int_E \rho\, \dot{u}'u\, dE = \frac{1}{2}\,\dot{a}'(t) M_{aa} \dot{a}(t)$$

(4.1.13) $$M_{aa} = \int_E \rho\, P'(x) P(x)\, dE$$

instead of being expressed in terms of a consistent mass matrix M_{aa} , can also be expressed in terms of consistent mass matrices for the boundary velocities and internal velocities:

(4.1.14) $$T = \frac{1}{2}\,\dot{q}'(t)\, M_{qq}\, \dot{q}(t) + \dot{q}'(t)\, M_{qb}\, \dot{b}(t) + \frac{1}{2}\,\dot{b}'(t) M_{bb} \dot{b}(t)$$

(4.1.15)
$$M_{qq} = \int_E \rho\, Q'(x)\, Q(x)\, dE = Q' M_{aa} Q$$
$$M_{qb} = \int_E \rho\, Q'(x)\, B(x)\, dE = Q' M_{aa}\, B = M'_{bq}$$
$$M_{bb} = \int_E \rho\, B'(x)\, B(x)\, dE = B' M_{aa} B$$

While

$$M_{aa} \quad \text{and} \quad \begin{pmatrix} M_{qq} & M_{qb} \\ M_{bq} & M_{bb} \end{pmatrix}$$

are certainly positive definite,

$$K_{aa} \text{ and } \begin{pmatrix} K_{qq} & K_{qb} \\ K_{bq} & K_{bb} \end{pmatrix}$$

are only non negative because the rigid body modes of the element, which must essentially be included in (4.1.1), are not associated with any strain energy.

Discretization of the external potential energy introduces a natural definition of the "generalized loads"

$$\int_{\partial E} u' \bar{t} \, d\partial E = q' g^* \tag{4.1.16}$$

where

$$g^* = \int_{\partial E} Q(x) \, \bar{t} \, d\partial E \tag{4.1.17}$$

There are no generalized loads conjugate to the bubble coordinates b , because the bubble functions precisely vanish along ∂E . The starred notation for the generalized loads conjugate to q indicates that they are linear functionals of the surface traction distribution and consequently provide only "weak" information about this distribution.

The functional (2.4.9) of the displacement principle can now be dispalyed in its completely discretized form for the element as

$$\int_{t_1}^{t_2} \frac{1}{2}\Big\{q'(K_{qq}q + K_{qb}b) + b'(K_{bq}q + K_{bb}b)$$

$$- \dot{q}'(M_{qq}\dot{q} + M_{qb}\dot{b}) - \dot{b}'(M_{bq}\dot{q} + M_{bb}\dot{b})\Big\} dt$$

$$- \int_{t_1}^{t_2} q' g^* dt .$$

(4.1.18)

Its variational derivatives with respect to q and b are

$$K_{qq}q + K_{qb}b + M_{qq}\ddot{q} + M_{qb}\ddot{b} = g^* \tag{4.1.19}$$

$$K_{bq}q + K_{bb}b + M_{bq}\ddot{q} + M_{bb}\ddot{b} = 0 . \tag{4.1.20}$$

In view of their contribution to the kinetic energy, the bubble coordinates are not as easily eliminated as in statics, where we can always write

$$b = - K_{bb}^{-1} K_{bq} q \tag{4.1.21}$$

because bubble functions necessarily represent independent deformation modes of the element and involve a positive definite K_{bb} matrix. If however, as is generally the case, we are mainly interested in the low frequency dynamic behaviour of the structure, we are entitled to assume that the bubble coordinates follow statically the boundary motions as in (4.1.21). Then, sustituting (4.1.21), not in equation (4.1.19), but directly into the kinetic and potential energies, we obtain equations of motion of type

$$K_E\, q_E + M_E \ddot{q}_E = g_E^* \qquad (4.1.22)$$

with "reduced" stiffness and mass matrices [21] :

$$K_E = K_{qq} - K_{qb} K_{bb}^{-1} K_{bq} \qquad (4.1.23)$$

$$M_E = M_{qq} - M_{qb} K_{bb}^{-1} K_{bq} - K_{bq} K_{bb}^{-1} M_{bq} + K_{qb} K_{bb}^{-1} M_{bb} K_{bb}^{-1} K_{bq}$$

This procedure is a particular case of a general reduction method to be presented in section 7. In (4.1.22) q_E and g_E^* stand respectively for q and g^*. When the bubble coordinates are not eliminated we still consider (4.1.22) as representing the equations of motion of the element, but here

$$q_E' = (q' \quad b') \qquad g_E^{*\prime} = (g^{*\prime} \quad 0)$$

and

$$K_E = \begin{pmatrix} K_{qq} & K_{qb} \\ K_{bq} & K_{bb} \end{pmatrix} \qquad M_E = \begin{pmatrix} M_{qq} & M_{qb} \\ M_{bq} & M_{bb} \end{pmatrix}$$

4.2 Structural stiffness and mass matrices

The principle of assembling the elements consists in stating that corresponding boundary displacements should have common interface values, implementing as shown earlier the exact transition conditions

$$u_+ = u_- \qquad \text{along interfaces}$$

If w denotes the column matrix of all independent generalized displacements at the structural level, the identification of displacements is achieved by means of Boolean, or incidence, matrices L_E (often called the element localizing matrix) addressing the elements of q_E to the proper ones of w:

$$q_E = L_E w \tag{4.2.1}$$

Equating the sum of the virtual work of all external forces acting on each element to the virtual work performed by the forces y^* conjugate to w , which are external to the assembled structure

$$\sum_E q'_E g^*_E = w' \sum_E L'_E g^*_E = w' y^*$$

and observing that this must hold for arbitrary w :

$$y^* = \sum_E L'_E g^*_E \tag{4.2.2}$$

Substitution of (4.1.22) yields finally

$$K w + M \ddot{w} = y^{*} \tag{4.2.3}$$

with

$$K = \sum_{E} L'_{E} K_{E} L_{E} \tag{4.2.4}$$

$$M = \sum_{E} L'_{E} M_{E} L_{E} \tag{4.2.5}$$

4.3 Dependent boundary displacements. Superelements

We turn to the case where equation (4.1.2), relating the time dependent coefficients of $a(t)$ and the boundary displacements $q(t)$ required to implement correctly the transition conditions, is such that

$M'g = 0$ admits non trivial solution

Let $g = Yc$ with arbitrary c denote its general solution (the columns of Y are independent). Then

$$Y'q = 0 \tag{4.3.1}$$

is a necessary and sufficient condition for the inversion of (4.1.2) and the boundary coordinates q are no more independent. This case typically presents itself in trying to set up conforming plate bending elements respecting the Kirchhoff-Love assumption of zero transverse strain [9] , [33] .

It can be solved by constructing a superelement, that is assembling a small number of elements, each of which suffers from dependency relations between its boundary displacements, in such a way that independence is obtained for the boundary displacements at the assembled level. Suppose that the coordinates of each component element are adressed either to the boundary coordinates $q_{(b)}$ of the superelement or to its interface coordinates $q_{(i)}$

$$q_E = F_E \, q_{(b)} + G_E \, q_{(i)} \tag{4.3.2}$$

and consider the dependency relations

$$Y'_E q_E = (Y'_E F_E) \, q_{(b)} + (Y'_E G_E) \, q_{(i)} = 0 \qquad E = 1,2, \ldots, N \tag{4.3.3}$$

Then, the complete set of dependency relations

$$F q_{(b)} + G q_{(i)} = 0 \qquad F = \begin{pmatrix} Y'_1 F_1 \\ \hline Y'_N F_N \end{pmatrix} \quad G = \begin{pmatrix} Y'_1 G_1 \\ \hline Y'_N G_N \end{pmatrix} \tag{4.3.4}$$

must be solvable for $q_{(i)}$. Thus the rows of G must be linearly independent. Then we can express

$$q_{(i)} = W q_{(b)} + B b \qquad (4.3.5)$$

where the columns of B represent possible "assembled" bubble modes for the superelement. From the inversion properties

$$F + GN = 0 \quad \text{and} \quad GB = 0 \qquad (4.3.6)$$

we deduce

$$Y'_E F_E + Y'_E G_E W = 0 \quad \text{and} \quad Y'_E G_E B = 0 \qquad (4.3.7)$$
$$E = 1,2, \ldots, N$$

so that

$$q_E = (F_E + G_E W) q_{(b)} + G_E B b \qquad E = 1,2, \ldots, N \quad (4.3.8)$$

obviously satisfies the dependency relation (4.3.3). A particular solution of

$$M_E a_E = (F_E + G_E W) q_{(b)} + G_E B b$$

is thus available for arbitrary $q_{(b)}$ and b

$$a_E = Q_E q_{(b)} + P_E b + B_E b_E \qquad (4.3.9)$$

the last term containing the eventual bubble modes of the component element.

The situation is now comparable to that of a simple element. The only difference is the necessity of extending the integrals required to compute stiffness and mass to the union of the domain E , or to the boundary of this union.

4.4 The equilibrium approach. Element flexibility and inverse-mass matrices [12], [13], [21].

The impulse field is discretized as follows

(4.4.1) $$\tau(x,t) = R(x)\,c(t) + S(x)\,s(t)$$

The stress distributions adopted are thus divided in two types:

a) In $S(x)$ each column is a set of stresses in equilibrium without body loads. Thus

(4.4.2) $$D'S(x) = 0$$

b) In $R(x)$ each column corresponds to a stress distribution that requires a non zero distribution of body loads

(4.4.3) $$D'R(x) = v(x)$$

In general $R(x)$ and $S(x)$ together constitute a complete representation of $\tau(x,t)$ in terms of polynomials up to a given degree and $v(x)$ contains at least the inertia loading due to a rigid body motion of the element.

The intensities $c(t)$ and $s(t)$ of the stress distributions are the unknown time functions to determine so that deformation compatibility be satisfied in some "best" sense.

Extremization of the functional (2.4.10) is a convenient tool to use for that purpose.

Along each facet $\partial_\alpha E$ of the boundary of an element, the assumption (4.4.1) generates a set of independent surface traction modes. To each surface traction mode a time dependent intensity is attached. The set of intensities constitutes a column vector $g_\alpha(t)$. Hence if N_α denotes the direction cosines operator for $\partial_\alpha E$ and $T_\alpha(x)$ the identified surface traction modes, the following identity must hold for arbitrary $\dot{c}(t)$ and $\dot{s}(t)$

$$N'_\alpha R(x)\dot{c} + N'_\alpha S(x)\dot{s} \equiv T_\alpha(x) g_\alpha(t) \qquad \text{on} \quad \partial_\alpha T \quad (4.4.4)$$

The elements of $g_\alpha(t)$ are called generalized loads; in contrast to the kinematical approach they give strong information about the surface tractions distribution, that can be reconstitued from the knowledge of their values. Each time a choice is made for the measure of intensity of a surface traction mode, this measure becomes related to the values of $\dot{c}$ and $\dot{s}$. Consequently we dispose of matrix relations

$$g_\alpha(t) = G_\alpha \dot{c}(t) + C_\alpha \dot{s}(t) \qquad \text{for } \partial_\alpha E$$

Denoting by $g(t)$ the column of all independent $g_\alpha(t)$ defined on the partial boundaries, we finally obtain

(4.4.5) $$g(t) = G\,\dot{c}(t) + C\,\dot{s}(t)$$

and this equation plays a role very similar to (4.1.2) in the kinematical models. The matrices G and C are called the load connexion matrices.

The virtual work of surface tractions on prescribed boundary displacements can now be subjected to a discretization coherent with (4.4.1) and (4.4.4). On the partial boundary $\partial_\alpha E$

(4.4.6) $$\int_{\partial_\alpha E} \bar{u}' t \, d\partial E = \int_{\partial_\alpha E} \bar{u}' \, T_\alpha(x)\, g_\alpha(t)\, d\partial E = q_\alpha^{*\prime} g_\alpha(t)$$

where

(4.4.7) $$q_\alpha^* = \int_{\partial_\alpha E} T'(x)\, \bar{u}\, d\partial E$$

Again, collecting in $q^*(t)$ the independent $q_\alpha^*(t)$ defined on partial boundaries, the virtual work of surface tractions receives a canonical scalar product form

(4.4.8) $$\int_{\partial E} \bar{u}' t \, d\partial E = g'(t)\, q^*(t)$$

The quantities defined by (4.4.7) are linear functionals of the boundary displacement field and provide only a weak knowledge of this field. It will be observed that from the viewpoint of strong and weak knowledge the roles of displacements and forces is here reversed as compared to the kinematical models.

We are now in a position to discretize completely the complementary variational principle (2.4.10):

$$\frac{1}{2}\int_E \dot{\tau}' H^{-1} \dot{\tau}\, dE = \frac{1}{2}\dot{c}' F_{cc} \dot{c} + \dot{c}' F_{cs} \dot{s} + \frac{1}{2}\dot{s} F_{ss} \dot{s} \qquad (4.4.9)$$

with a positive definite flexibility matrix

$$F = \begin{pmatrix} F_{cc} & F_{cs} \\ F_{sc} & F_{ss} \end{pmatrix} \qquad (4.4.10)$$

$$F_{cc} = \int_E R'(x)\, H^{-1}\, R(x)\, dE$$

$$F_{cs} = \int_E R'(x)\, H^{-1}\, S(x)\, dE$$

$$F_{ss} = \int_E S'(x)\, H^{-1}\, S(x)\, dE$$

and

$$\frac{1}{2}\int_E \frac{1}{\rho}(D'\tau)'(D'\tau)\, dE = \frac{1}{2} c' N c \qquad (4.4.11)$$

which introduces a positive definite "inverse-mass" matrix

$$N = \int_E \frac{1}{\rho}\left(D'R(x)\right)' D'R(x)\, dE = \int_E \frac{1}{\rho}\, v'(x)\, v(x)\, dE \qquad (4.4.12)$$

Substituting (4.4.9), (4.4.11) and (4.4.8) into the variational principle (2.4.10), the variational derivatives, with δc and δs vanishing at the time limits, yield

$$(4.4.13) \qquad -(F_{cc}\ddot{c} + F_{cs}\ddot{s}) - Nc + G'\dot{q}^* = 0$$

$$(4.4.12) \qquad -(F_{sc}\ddot{c} + F_{ss}\ddot{s}) \qquad + C'\dot{q}^* = 0$$

They are the compatibility conditions looked for.

4.5 Solution of equilibrium approach in terms of unknown displacements

The use of unknown generalized displacements for assembling finite elements at the structural level by localization is a convenient procedure that is also applicable to equilibrium types of elements. The only difference is that the generalized displacements are "weak" and are essentially of the interface type. Hence interface identification of weak displacements will result in equilibrium of conjugate generalized loads. Since those however are "strong" they will enforce complete interface equilibrium of surface tractions, which was a requirement of the complementary variational principle at the structural level. The weak displacements on the boundary, $q^*(t)$, were already defined but we have still to attach weak conjugate to the body loads generated by (4.4.3). This is again obtained by virtual work considerations. The body loads are given by

$$- D'\dot{\tau} = - D'R(x)\,\dot{c} = - v(x)\,\dot{c} \qquad (4.5.1)$$

Their virtual work, put in canonical form,

$$- \int_E u'D'\dot{\tau}\,dE = p^{*\prime}\dot{c} \qquad (4.5.2)$$

yields

$$p^* = - \int_E v'(x)\,u\,dE \qquad (4.5.3)$$

linear functionals of the internal displacement field.
Now $- D'\dot{\tau}$ also represents the inertia forces $- \rho\,\ddot{u}$. Consequently, considering the kinetic energy,

$$\frac{d}{dt}\frac{1}{2}c'Nc = \dot{c}'Nc = \int_E (\rho\,\ddot{u})'\,\dot{u}\,dE = \int_E (D'\dot{R})'\dot{u}\,dE$$

$$= \dot{c}' \int_E v'(x)\,\dot{u}\,d\tau = - \dot{c}'\dot{p}^*$$

and, comparing,

$$\dot{p}^* = - Nc \qquad (4.5.4)$$

The introduction of p^* as conjugate to $\dot{c}$ also furnishes an interpretation to

$$c = - N^{-1}\dot{p}^* \qquad (4.5.5)$$

In view of (4.4.18) the compatibility equation (4.4.13) and (4.4.14) can be written together as

$$(4.5.6) \qquad F_E \begin{pmatrix} \ddot{c} \\ \ddot{\jmath} \end{pmatrix}_E = C'_E \begin{pmatrix} \dot{q}^* \\ \dot{p}^* \end{pmatrix}_E$$

in terms of the elements complete flexibility matrix (4.4.10) and a complete load connexion matrix

$$(4.5.7) \qquad C_E = \begin{pmatrix} G & C \\ I & 0 \end{pmatrix}$$

Noting that (4.4.5) can be written in the form

$$(4.5.8) \qquad C_E \begin{pmatrix} \dot{c} \\ \dot{\jmath} \end{pmatrix}_E = \begin{pmatrix} g \\ \dot{c} \end{pmatrix}_E$$

differentiating and substituting in this the solution of (4.5.6), we find

$$(4.5.9) \qquad \begin{pmatrix} \dot{g} \\ \ddot{c} \end{pmatrix}_E = K_E \begin{pmatrix} \dot{q}^* \\ \dot{p}^* \end{pmatrix}_E$$

with

$$(4.5.10) \qquad K_E = C_E F_E^{-1} C'_E$$

the stiffness matrix of the equilibrium element. Introducing now the mass matrix of the element, defined as

$$(4.5.11) \qquad M_E = \begin{pmatrix} 0 & 0 \\ 0 & N^{-1} \end{pmatrix}$$

The system (4.5.9) is finally presented in the same form as that (4.1.26) of a kinematical element

$$K_E \begin{pmatrix} \dot{q}^* \\ \dot{p}^* \end{pmatrix}_E + M_E \begin{pmatrix} \dddot{q}^* \\ \dddot{p}^* \end{pmatrix}_E = \begin{pmatrix} \dot{g} \\ 0 \end{pmatrix}_E \qquad (4.5.12)$$

However, while in the former case it was a discretized form of the dynamic equilibrium equations, in the present case it is a discretized form of the compatibility conditions (2.4.5).

4.6 Kinematical freedoms of equilibrium elements

A set of weak displacements

$$\begin{pmatrix} q^* \\ p^* \end{pmatrix} = u^*$$

that produces no strain energy is one for which

$$u^{*\prime} K_E u^* = (C'_E u^*)' F^{-1} (C'_E u^*) = 0$$

Since F is positive definite, such weak displacements are all found as non trivial solutions of the homogeneous system

$$C'_E u^* = 0 \qquad (4.6.1)$$

or, in view of the structure (4.4.21) of C_E

$$C' q^* = 0 \qquad\qquad p^* = - G' q^* \qquad (4.6.2)$$

Attention can thus be focused on the first of equations (4.6.2). For any non trivial solution it furnishes for the boundary part

q^* , the second equation furnishes the corresponding internal part p^* . Rigid body modes of the element must evidently produce solutions. They can be found by inserting rigid body displacements fields for $\bar{u}$ into the definitions (4.4.7). Any non trivial solution other than rigid body modes is a kinematical freedom of the element; it is an undesirable feature, the exact converse of the non independence of generalized boundary displacements in kinematical models. Indeed it can be viewd as an undesirable dependence constraint on the generalized loads. Equations (4.6.1) is he homogeneous adjoint of (4.5.8) and provides the necessary and sufficient conditions for (4.5.8) to be invertible. More simply, the non trivial solutions q^* of the first of equations (4.6.2) provide the conditions

(4.6.3) $$u^{*\prime}(g - G\dot{c}) = 0$$

for $\dot{\jmath}$ to be expressible in terms of g and $\dot{c}$ in (4.4.5).

As long as u^* represents a rigid body mode, the conditions (4.6.3) merely express the required global equilibrium between forces applied at the surface of the element and inertia loads. But if there are additional kinematical freedoms the set of g forces is submitted to further restrictions. It is also interesting to note in (4.4.5) the possible existence of non trivial $\dot{\jmath} = h$ vectors, such that

(4.6.4) $$C\,h = 0$$

The corresponding states of stress

$$\dot{\tau} = S(x)\,h \tag{4.6.5}$$

do not produce surface tractions and are in equilibrium without body forces. They are really self-stressing states within the element. By analogy with the kinematical models they could be called stress bubble modes.

4.7 Structural assembling of equilibrium models

Turning back to equation (4.5.9) after partionning the stiffness matrix

$$\dot{g}_E = K_{qqE}\,\dot{q}^*_E + K_{qpE}\,\dot{p}^*_E \tag{4.7.1}$$

$$\ddot{c}_E = K_{pqE}\,\dot{q}^*_E + K_{ppE}\,\dot{p}^*_E \tag{4.7.2}$$

we substitute

$$c_E = -\,N_E^{-1}\,\dot{p}^*_E \tag{4.7.3}$$

and localize the weak boundary displacements by

$$q^*_E = L_E\,w^* \tag{4.7.4}$$

From (4.7.2) we then obtain a first set of dynamic equations

(4.7.5) $$-N_E^{-1}\dddot{p}_E^{*} = K_{pqE}L_E\dot{w}^{*} + K_{ppE}\dot{p}_E^{*} \qquad E = 1,2\ldots N$$

If there are no interface and external boundary inertia loads

(4.7.6) $$\sum L'_E g_E = 0$$

Time differentiation of this and substitution of (4.7.1) and (4.7.4) produces

(4.7.7) $$\left(\sum_E L'_E K_{EE} L_E\right)\dot{w}^{*} + \sum_E L'_E K_{qqE}\dot{p}_E^{*} = 0$$

The structural eigenvalue problem is finally reduced to the standard form

(4.7.8) $$(\mathring{K} - \omega^2\mathring{M})\mathring{x} = 0$$

where $\mathring{x}$ is a vector of amplitudes of the weak displacements $p_1^{*}, p_2^{*}, \ldots p_N^{*}, w^{*}$ and

$$\mathring{K} = \begin{bmatrix} K_{pp1} & 0 & 0 & 0 & K_{pq1}L_1 \\ 0 & K_{pp2} & 0 & 0 & K_{pq2}L_2 \\ 0 & 0 & \ddots & 0 & 0 \\ 0 & 0 & 0 & K_{ppN} & K_{pqN}L_N \\ L'_1K_{qp1} & L'_2K_{qp2} & 0 & L'_NK_{qpN} & \sum_E L'_E K_{qqE} L_E \end{bmatrix}$$

$$\mathring{M} = \begin{bmatrix} N_1^{-1} & 0 & 0 & 0 & 0 \\ 0 & N_2^{-1} & 0 & 0 & 0 \\ 0 & 0 & \ddots & 0 & 0 \\ 0 & 0 & 0 & N_N^{-1} & 0 \\ 0 & 0 & 0 & 0 & 0 \end{bmatrix}$$

Because of the absence of any inertia attached to the w^* part of the displacements (the last row and column of $\mathring{M}$ are zero), it is possible to eliminate w^* statically by using the equation (4.7.7). The degrees of freedom of the eigenvalue problem are thus reduced to the set of all internal degrees of freedom p_E^* . Because this number is usually quite low compared to a kinematical approach of equivalent idealization, the equilibrium approach to dynamics appears as a natural "eigenvalue economizer".

5. EIGENVALUE ANALYSIS IN THE PRESENCE OF KINEMATICAL MODES

5.I Introduction

The problem of determining the natural frequencies and mode shapes of structures by matrix iteration on the linear system

(5.1.1) $$K x = \omega^2 M x$$

becomes complicated when the stiffness matrix K , instead of being positive definite, is only non negative. This situation prevails in free-free beams, for instance, or, more generally, whenever the structure is capable of undergoing displacement modes without storing deformation energy. Such modes will be indifferently referred to as kinematical or rigid body modes. The matrix K being singular, a commonly proposed procedure [3] consists in applying a spectral shift α and solve the modified problem

(5.1.2) $$(K + \alpha M) x = (\omega^2 + \alpha) M x$$

In an improved version of this procedure the spectral shift is obtained by separate elementary modifications to the stiffness and mass matrices in order to eliminate the presence of the kinematical modes [14] .

The technique discussed here consists in generalizing the concept of a flexibility matrix to positive semi-definite stiffness matrices. It was first proposed in [8] and can be modified in order to obtain a symmetrical iteration matrix. When applied to an analysis in finite elements, it requires the non-trivial preliminary operation of determining the kinematical modes. This can be achieved by applying the Gauss inversion algorithm to the original stiffness matrix. Such a numerical procedure, using a Choleski factorization, was first applied by Craig and Bampton [5] . However advantage can be gained from the Gaussian inversion ability to preserve the symmetry of the original matrix while selecting at each stage the largest pivot on the diagonal. Moreover the latter procedure is characterized by a minimal growth of round-off errors [39] , [40] .

5.2 Kinematical modes and deformation modes

K is positive but only semi-definite if there exists non trivial solutions of

$$K x = 0 \tag{5.2.1}$$

They may obviously be considered as modal shapes of problem (5.1.1) associated to a zero eigenvalue. We denote a fundamental set (linearly independent and complete) of such solutions by

$x = u_{(i)}$ $(i = 1,2 \ldots \rho)$ and refer to them as the kinematical modes. The other eigensolutions of problem (5.1.1) are modal shapes associated to non zero eigenvalues. A complete set of those will be denoted by $x = x_{(r)} (r = 1,2 \ldots \sigma)$; they are referred to as the deformation modes.

Modal shapes belonging to different eigenvalues are known to be orthogonal with respect to both the stiffness and the mass matrix. Modes belonging to the same eigenvalue can always be rendered orthogonal to one another. Consequently and without loss of generality

$$u'_{(i)} M u_{(j)} = \delta_{ij} \qquad x'_{(s)} M x_{(r)} = \delta_{sr}$$

(5.2.2)

$$u'_{(i)} M x_{(r)} = 0 \qquad x'_{(s)} K x_{(r)} = \omega_r^2 \delta_{rs}$$

Relations like $x'_{(r)} K u_{(i)} = 0$ and $u'_{(j)} K u_{(i)} = 0$ are only weak consequences of the equation

(5.2.3) $$K u_{(i)} = 0$$

satisfied by the kinematical modes. Equations (5.2.2) show additionally that the modal shapes were normed to unity with respect to the mass matrix.

Introduction of the modal matrices

$$U = \left\{ u_{(1)} \ldots u_{(\rho)} \right\} \qquad X = \left\{ x_{(1)} \ldots x_{(\sigma)} \right\}$$

allows to present the same information in matrix form

$$KU = 0 \tag{5.2.4}$$

$$U'MU = E_\rho \tag{5.2.5}$$

$$X'MX = E_\sigma \tag{5.2.6}$$

$$U'MX = X'MU = 0 \tag{5.2.7}$$

$$X'KX = \Delta \tag{5.2.8}$$

where E_ρ and E_σ are identity matrices of dimensions $\rho \times \rho$ and $\sigma \times \sigma$ respectively, while Δ is a diagonal matrix of the eigenvalues ω_r^2 .

$(U \quad X)$ being a base matrix (square with linearly independent columns) any vector x admits of a unique modal expansion

$$x = U a + X b \tag{5.2.9}$$

whose columns of coefficients are immediatly obtained by application of the orthogonality relations as

$$a = U'M x \qquad b = X'M b \tag{5.2.10}$$

Similarly, M being positive definite, $(MU \quad MX)$ is a natural base matrix for the unique expansion of an arbitrary force amplitude vector

$$p = MUc + MXd \tag{5.2.11}$$

with

$$c = U'p \qquad\qquad d = X'p \tag{5.2.12}$$

The modal expansion (5.2.11) is in terms of the inertia force distribution pertaining to each modal shape.

5.3 Static equilibrium conditions

Let us momentarily turn to the more general equations of motion

$$Kq + M\ddot{q} = p(t) \tag{5.3.1}$$

from which problem (5.1.1) is derived by setting $p(t) = 0$ and assuming harmonic free vibrations ($q = x \cos \omega t$). Introducing normal coordinates

$$q(t) = \sum_1^{\rho} \eta_i(t)\, u_{(i)} + \sum_1^{\sigma} \xi_r(t)\, x_{(r)} \tag{5.3.2}$$

equations (5.3.1) are transformed in a set of uncoupled single degree of freedom oscillators, governed by the "normal equations"

$$\ddot{\eta}_i = u'_{(i)} p(t) \qquad (i = 1, \ldots, \rho) \qquad (5.3.3)$$

$$\omega_r^2 \xi_r + \ddot{\xi}_r = x'_{(r)} p(t) \qquad (r = 1, \ldots, \sigma) \qquad (5.3.4)$$

If the external load vector p is static (independent of time) the normal equations show clearly that the necessary and sufficient condition for the existence of a static response are

$$u'_{(i)} p = 0 \qquad (i = 1, \ldots, \rho) \qquad (5.3.5)$$

Each condition expresses the nullity of the virtual work performed by the static load on a kinematical displacement mode. They are a reminder of the fundamental theorem of statics through virtual work and are fully equivalent to the conditions of global equilibrium.

5.4 The projection operator A. Pseudo-inversion of K.

The static equilibrium conditions (5.3.5), that can be summarized in the matrix equation

$$U' p = 0 \qquad (5.4.1)$$

show, by reference to (5.2.12), that the modal expansion of the static load p should not contain the part pertaining to the inertia loads of the kinematical modes ($c = 0$).

Then, to any arbitrary loading vector p corresponds a modified loading vector $\hat{p}$ satisfying the global equilibrium conditions (5.3.1)

$$\hat{p} = p - MUc = p - MUU'p$$

or

(5.4.2) $$\hat{p} = A'p$$

with

(5.4.3) $$A' = E - MUU'$$

(E denotes here the identity matrix for the complete vector space of dimensions $\rho + \sigma$).

As a verification we find

(5.4.4) $$U'\hat{p} = U'A'p = 0$$

because

(5.4.5) $$U'A' = U' - (U'MU)\,U' = U' - U' = 0$$

Moreover we find

(5.4.6) $$X'\hat{p} = X'A'p = X'p$$

because

(5.4.7) $$X'A' = X' - (X'MU)\,U' = X'$$

(5.4.5) and (5.4.7) are equivalent to the selective properties of the projector operator A :

$$AU = 0 \quad \text{or} \quad A u_{(i)} = 0 \quad (i = 1, \ldots, \rho) \tag{5.4.8}$$

$$AX = X \quad \text{or} \quad A x_{(r)} = 0 \quad (r = 1, \ldots, \sigma) \tag{5.4.9}$$

Moreover, we have.

$$A' M U = 0 \quad \text{or} \quad A' M u_{(i)} = 0 \tag{5.4.10}$$

$$A \; M X = M X \quad \text{or} \quad A' M x_{(r)} = M x_{(r)} \tag{5.4.11}$$

The relationship (5.4.2) has a well known dynamical interpretation that becomes apparent when it is reformulated as

$$\hat{p} = p - \sum_i (u'_{(i)} p) M u_{(i)}$$

and use is made of the dynamic equations (5.3.3)

$$\hat{p} = p - \sum_i \eta_i M u_{(i)}$$

The application of the arbitrary static loading p induces accelerations in the kinematical degrees of freedom. Suppressing the deformation modes, the addition of the induced inertia loadings to p , as in the right-hand side, yields according to d'Alembert's principle a self-equilibrated system.

By introducing (5.4.2) we have now insured the existence of a solution to the modified static problem

(5.4.12) $$K q = A' p$$

for arbitrary p . The solution is however not unique; to any particular solution $q = F_o p$, we can add a general solution to the corresponding homogeneous equation. Thus

(5.4.13) $$q = F_o p + U g$$

where g is arbitrary. As any g can always be represented in the form

$$g = R' p$$

by a suitable matrix R' , the general solution can be placed in the form

(5.4.14) $$q = (F_o + U R') p$$

This solution is a generalized pseudo-inverse of the singular static problem

(5.4.15) $$K p = p$$

(5.4.16) $$F = F_o + U R'$$

is a flexibility matrix that is a generalized pseudo-inverse of the singular stiffness matrix

In [8] it was called a matrix of extended influence coefficients. Substituting

$$q = F p \tag{5.4.17}$$

into (5.4.12) and noting that p is arbitrary

$$K F = A' \tag{5.4.18}$$

This pseudo-inverse relationship replaces the classical inversion

$$K K^{-1} = E$$

valid in the absence of kinematical freedoms.

5.5 Unicity of a pseudo-inverse. Isostaticity constraints.

The whole problem of finding a pseudo-inverse to K consists in obtaining a particular solution F to equations (5.4.12). The practical answer to this problem is precisely related to the introduction of additional requirements that remove the arbitrariness in F. Suppose that we require of the pseudo-inversion (5.4.17) that the solution q be orthogonal to the kinematical modes for arbitrary p :

$$U' M q = 0 \tag{5.5.1}$$

Then, substituting (5.4.13) and using (5.2.5)

$$U'MF_o p + g = 0$$

hence

$$q = (F_o - UU'MF_o)\ p$$

This solution of (5.4.12) will be denoted by

(5.5.2) $$q = Gp$$

and the pseudo - inverse

(5.5.3) $$G = F_o - UU'MF_o = AF_o$$

will be shown to be independent of the particular choice F_o. Indeed it is readily apparent from (4.16) that any other particular choice F_1 is related to F_o by

$$F_1 = F_o + UR'$$

where R is some fixed matrix. But then, in view of (5.4.8)

$$AF_1 = AF_o$$

The pseudo - inverse G is not only unique, it is also symmetrical. A simple proof consists in transposing relation (5.4.18) and post multiplying it by F , then

$$F'KF = AF = G$$

and in view of the structure of the left-hand side, symmetry is obvious. Except for the fact that orthogonality is here defined in terms of a positive definite matrix M and not the identity matrix, G coincides with the mathematical pseudo - inverse of K as defined by Penrose [31] . It can also be shown [8] to coincide with the spectral expansion of the dynamic coefficients matrix $(K - \omega^2 M)^{-1}$, amputated of the terms corresponding to the kinematical mode shapes:

$$G = \sum_{1}^{\sigma} \frac{1}{\omega_r^2} x_{(r)} x'_{(r)} \qquad (5.5.4)$$

The unicity of (5.4.17) follows more generally from a set of additional constraints of type

$$S'q = 0 \qquad (5.5.5)$$

provided $S' U$ be a non singular $\rho \times \rho$ matrix. Indeed, substituting (5.4.13) we must have

$$S' F_0 p + S' U g = 0$$

and then

$$g = -(S' U)^{-1} S' F_0 p$$

whereby

$$q = F_0 - U (S' U)^{-1} S' F_0 p$$

and a more general pseudo - inverse

(5.5.6) $$F = \left(E - U\,(S'\,U)^{-1}S'\right)F_0 = PF_0$$

Unicity is proved by the same technique

$$PF_1 = PF_0 + P\,U\,R' = PF_0$$

because

$$P\,U = \left(E - U\,(S'\,U)^{-1}S'\right)U = 0$$

Constraints such as (5.5.5) can be qualified by their property to suppress the kinematical freedoms; for if we restrict displacements to the kinematical modes

$$q = U\,a$$

and apply the constraints, there follows

$$S'U\,a = 0 \quad , \text{ implying} \quad a = 0$$

since $S'\,U$ is non singular. Consequently we can conceive of (5.5.5) as physical constraints linking the structure to a solid reference frame in an isostatic manner. The structure can then accept an arbitrary static loading p and a unique flexibility or influence coefficients matrix G_{iso} can be produced such that

$$q = G_{iso}\,p$$

G_{iso} is symmetrical and has for instance the elementary structure

$$G_{iso} = \begin{pmatrix} K_{\sigma\sigma}^{-1} & 0 \\ 0 & 0 \end{pmatrix} \qquad (5.5.7)$$

when assuming that suitable constraints are obtained by preventing the ρ last displacement components to take place. An automated selection of the constraints will be presented in section 5.7. When the arbitrary loading p is replaced by $A'p$, the structure becomes self-equilibrated without the reaction loads due to the constraints. As a matter of fact, since they are isostatically determined, they can but vanish and we can conclude that

$$q = G_{iso} A' p$$

is a possible displacement vector of the unconstrained structure. The role played by the constraints is reduced to the determination of a particular set of displacements without the indeterminacy contained in (5.4.16); the unicity of pseudo - inverses of type (5.5.6) is thus physically clarified. Moreover the solution found

$$F = G_{iso} A' \qquad (5.5.8)$$

is now quite easily obtainable. It also furnishes a new direct proof of the symmetry of

(5.5.9) $$G = AF = AG_{iso}A'$$

5.6 Iteration in semi-definite eigenvalue problems

The deformation modes $x_{(r)}$, which satisfy

$$K\, x_{(r)} = \omega_r^2\, M\, x_{(r)}$$

will also satisfy the system

$$F'K\, x_{(r)} = \omega_r^2\, F'M\, x_{(r)}$$

where F' is the transpose of a pseudo - inverse of K . Considering the transpose of the property (5.4.18) and using (5.4.9), this can be written in the form

(5.6.1) $$F'M\, x_{(r)} = \lambda_r\, x_{(r)}$$

(5.6.2) $$\lambda_r = 1/\omega_r^2$$

and suggests the use of $F'M$ in a power iteration scheme to solve the new eigenvalue problem

(5.6.3) $$F'M\, x = \lambda\, x$$

Because of (5.6.1) this problem admits precisely the unknown deformation modes as eigensolutions. There remains however to investigate the nature of the ρ still missing eigensolutions. They can be termed "parasitic solutions" because the kinematical modes, though eigensolutions of the original problem (5.1.1), are generally no more solutions of (5.6.3). It is a remarkable property of pseudo - inverses F obtained through the use of isostaticity constraints that they induce in (5.6.3) parasitic solutions of zero characteristic value λ, which are thus wiped out at the very first iteration. To this purpose we must prove that

$$F' M w = 0 \tag{5.6.4}$$

admits ρ non trivial linearly independent solutions. Indeed, when the construction of F is based on (5.5.5), we have obviously

$$S' F = 0 \qquad \text{or} \qquad F' S = 0 \tag{5.6.5}$$

a property that is also easily verified on (5.5.6). Thus

$$w = M^{-1} S s \tag{5.6.6}$$

is a non trivial solution of (5.6.4) for arbitrary s and, the ρ columns of S being linearly independent the proof is completed. Observe that when F is taken to be symmetrical G matrix, corresponding to the constraints (5.5.1), the parasitic modes

$$w = M^{-1}(U'M)' s = U s$$

are the kinematical modes themselves, whose natural frequency zero has thus been shifted to infinity. As a final conclusion the problem (5.6.3) in which F is of type (5.5.8) or

$$A\,G_{iso}\,M x = \lambda x \tag{5.6.7}$$

is well adapted to power iteration and will provide initial convergence towards the deformation mode of largest characteristic number λ (or smallest natural frequency). Classical deflation algorithms [11] , [21] are applicable to the successive determination of modes of higher frequency.

5.7 Numerical computation of kinematical modes

Suppose that rows and columns of the original stiffness matrix are ordered in such a manner that suppression of the ρ last degrees of freedom, q_ρ , suppresses the kinematical degrees of freedom. The other $\sigma = n - \rho$ degrees of freedom may thus be interpreted as generalized deformation coordinates.

Hence the singular static problem (5.4.15) admits of the partitioned form

$$\begin{pmatrix} K_{\sigma\sigma} & K_{\sigma\rho} \\ K_{\rho\sigma} & K_{\rho\rho} \end{pmatrix} \begin{pmatrix} q_\sigma \\ q_\rho \end{pmatrix} = \begin{pmatrix} p_\sigma \\ p_\rho \end{pmatrix} ; \tag{5.7.1}$$

the static equilibrium condition to be verified by the right-hand side vector will be explicited further.

Performing σ Gaussian inversion steps leads to the set of equations

$$q_\sigma = K_{\sigma\sigma}^{-1}(p_\sigma - K_{\sigma\rho} q_\rho) \tag{5.7.2}$$

and substitution of this into the second equation (5.7.1) transforms the initial system as follows:

$$\begin{bmatrix} K_{\sigma\sigma}^{-1} & -K_{\sigma\sigma}^{-1} \cdot K_{\sigma\rho} \\ K_{\rho\sigma} \cdot K_{\sigma\sigma}^{-1} & K_{\rho\rho} - K_{\rho\sigma} \cdot K_{\sigma\sigma}^{-1} \cdot K_{\sigma\rho} \end{bmatrix} \begin{bmatrix} p_\sigma \\ q_\rho \end{bmatrix} = \begin{bmatrix} q_\sigma \\ p_\rho \end{bmatrix} . \tag{5.7.3}$$

From a computational point of view, it is essential to point out at this stage that the symmetry of the initial matrix will be preserved by the following changes of sign

$$\begin{bmatrix} -K_{\sigma\sigma}^{-1} & K_{\sigma\sigma}^{-1} \cdot K_{\sigma\rho} \\ K_{\rho\sigma} \cdot K_{\sigma\sigma}^{-1} & K_{\rho\rho} - K_{\rho\sigma} \cdot K_{\sigma\sigma}^{-1} \cdot K_{\sigma\rho} \end{bmatrix} \begin{bmatrix} -p_\sigma \\ -q_\rho \end{bmatrix} = \begin{bmatrix} q_\sigma \\ -p_\rho \end{bmatrix} . \tag{5.7.4}$$

Next consider the submatrix $K_{\rho\rho} - K_{\rho\sigma} \cdot K_{\sigma\sigma}^{-1} \cdot K_{\sigma\rho}$ which, in view of the positive semi-definite character of the initial matrix, is expected to vanish. The proof follows by setting the linear homogeneous problem (5.2.1) into the partitioned form

$$\begin{bmatrix} K_{\sigma\sigma} & K_{\sigma\rho} \\ K_{\rho\sigma} & K_{\rho\rho} \end{bmatrix} \begin{bmatrix} q_{\sigma} \\ q_{\rho} \end{bmatrix} = 0 , \tag{5.7.5}$$

the solution of which are the kinematical mode shapes. Indeed the first equation (5.7.5) associates to any arbitrary choice of the kinematical degrees of freedom q_{ρ}, the set of dependent generalized coordinates

$$q_{\sigma} = - K_{\sigma\sigma}^{-1} \cdot K_{\sigma\rho} q_{\rho} . \tag{5.7.6}$$

Introducing (5.7.6) into the second equation (5.7.5) yields

$$(K_{\rho\rho} - K_{\rho\sigma} \cdot K_{\sigma\sigma}^{-1} \cdot K_{\sigma\rho}) \; q_{\rho} = 0$$

for any q_{ρ}, terminating the proof.

The static problem (5.7.4) reduces thus to the form

$$\begin{bmatrix} -K_{\sigma\sigma}^{-1} & K_{\sigma\sigma}^{-1} \cdot K_{\sigma\rho} \\ K_{\rho\sigma} \cdot K_{\sigma\sigma} & 0 \end{bmatrix} \begin{bmatrix} -p_{\sigma} \\ -q_{\rho} \end{bmatrix} = \begin{bmatrix} q_{\sigma} \\ -p_{\rho} \end{bmatrix} , \tag{5.7.7}$$

from which one deduces the expected global equilibrium conditions

$$K_{\rho\sigma} \cdot K_{\sigma\sigma}^{-1} p_\sigma = p_\rho \ . \tag{5.7.8}$$

As another consequence of (5.7.6), the modal matrix

$$\begin{pmatrix} -K_{\sigma\sigma}^{-1} \cdot K_{\sigma\rho} \\ \\ E_\rho \end{pmatrix} \tag{5.7.9}$$

defines the linearly independent set of kinematical modes associated to a unit displacement of each kinematical degree of freedom successively. These modes still have to be orthonormed in order to verify the assumed properties (5.2.5).

On the other hand, setting the ρ arbitrary independent coordinates q_ρ equal to zero provides a particular isostatic reference frame. Hence (5.7.7) also yields

$$G_{iso} = \begin{pmatrix} K_{\sigma\sigma}^{-1} & 0 \\ \\ 0 & 0 \end{pmatrix} . \tag{5.7.10}$$

The numerical inversion procedure will thus be organized as follows:

1. The pivot element is chosen at each inversion step as the largest term of the diagonal which has not yet been utilized: say

$$k_{ii} = \max_j k_{jj} \qquad j = 1, \dots n \ . \tag{5.7.11}$$

2. The other elements are transformed according the following rules:

$$- \quad k_{ii}^{*} = -\frac{1}{k_{ii}} \tag{5.7.12}$$

$$- \quad k_{ij}^{*} = k_{ji}^{*} = \frac{k_{ij}}{k_{ii}} = \frac{k_{ji}}{k_{ii}} \qquad j \neq i \tag{5.7.13}$$

$$- \quad k_{j\ell}^{*} = k_{\ell j}^{*} = k_{j\ell} - \frac{k_{ji} \cdot k_{i\ell}}{k_{ii}} \qquad j, \ell \neq i$$

(5.7.14)

$$= k_{\ell j} - \frac{k_{\ell i} \cdot k_{ij}}{k_{ii}}$$

3. The procedure is restarted until all non zero terms on the diagonal have been inverted.

The symmetry of the initial stiffness matrix has obviously been preserved by the sequence of operations (5.7.11) to (5.7.14), and the algorithm may thus be performed even when limiting the memorization of the K matrix to its lower - or upper - half triangular part. Note that other procedures like Choleski factorization [5] or Gauss-Jordan elimination [21] do not permit the pivot selection anywhere on the diagonal when storing the initial matrix under triangular form. As a consequence of this symmetry property of the Gaussian inversion algorithm, the associated growth of round-off errors is minimized when choosing the best pivot element at each inversion step [39] , [40] .

At the end of the inversion procedure, the kinematical modes are obtained on rows and columns associated with the zero diagonal terms. Setting then equal to zero the same rows and columns gives the symmetric matrix G_{iso} . Both operations needed have thus been performed together.

5.8 Obtention of a symmetric iteration matrix

Consider first the change of variables

$$\bar{x} = T x \tag{5.8.1}$$

where T denotes the upper-triangular matrix resulting from the Choleski decomposition of the mass matrix:

$$M = T'T \tag{5.8.2}$$

The transformed kinematical modal matrices

$$\bar{U} = T U$$

verify the simpler orthonormal relations

$$\bar{U}'\bar{U} = E_{\rho} \tag{5.8.3}$$

and the eigenvalue problem

(5.8.4) $$A\, G_{iso}\, A'\, M x = \lambda x$$

has to be rewritten as

(5.8.5) $$T A\, G_{iso}\, A'\, T' \bar{x} = \lambda \bar{x}$$

If we still introduce the new projection operator

(5.8.6) $$\bar{A} = I - \bar{U}\, \bar{U}'$$

(5.8.5) changes into

(5.8.7) $$\bar{A}\, \bar{G}_{iso}\, \bar{A}'\, \bar{x} = \lambda \bar{x}$$

with the transformed flexibility matrix

(5.8.8) $$\bar{G}_{iso} = T\, G_{iso}\, T'$$

Note that the matrix product (5.8.8) does not involve a larger number of operations than (5.6.5) if taking into account the triangular form of T . Finally, the pre- and post-multiplication by the projection operator reduces the operation of obtaining the iteration matrix to a sum of dyadic products

$$S = \bar{A}\, \bar{G}_{iso}\, \bar{A}'$$

(5.8.9) $$= \bar{G}_{iso} - \bar{U}\, \bar{U}' \bar{G}_{iso} - \bar{G}_{iso}\, \bar{U}\, \bar{U}'$$

$$+ \bar{U} \cdot (\bar{U}'\, G_{iso} \bar{U}) \cdot \bar{U}'$$

The use of (5.8.9) extends to hypostatic structures the possibility of reducing the memorization of the dynamic flexibility matrix to its upper or lower triangular part [21] .

6. THE REDUCTION METHODS

6.I. Introduction

It is economically unrealistic to solve eigenvalue problems involving more than a few hundred degrees of freedom on present day computers. As acceptable finite element idealizations of aerospace and civil engineering structures generally involve thousands of degrees of freedom, it becomes necessary to reduce the size of the eigenvalue problems by methods that preserve the accuracy of the low frequency spectrum.

A first approach outlined in this paper uses the classical elimination [27] , [43] based on the assumption that a certain number of degrees of freedom may be reduced by static considerations. Use was made of the Kato and Temple theorems [28] , [37] on bounding of eigenvalues to improve the algorithms by adding the possibility of computing bounds to the error introduced by condensation [17] . Coupled with the finite element method, the elimination algorithm allows an easy step build-up of the whole structure (substructures in series). The results obtained in treating large scale applications show that the loss of accuracy for the lower frequency modes is negligible.

An alternative, briefly discussed in this paper, concerns the method of coupling substructures in parallel and consists in treating the whole structure as an array of several sub-regions already separately condensed.

In both cases it is agreed to limit either the deformation modes of the structure, or those of its constitutive parts, to some known modes, $r_{(i)}$, smaller in number than the one n of the degrees of freedom of the initial idealization. They are chosen as the most representative of the low frequency behaviour of the structure.

The possible motions of the structure, q , are thus restricted to the form

$$q = R\,a \tag{6.1.1}$$

where R denotes the $(n \times m)$ matrix collecting the selected modes $r_{(i)}$.

Expressing that the potential energies associated with a displacement mode are restrained by (1.1) gives

$$U = \frac{1}{2}\, a'(R' K R)\, a \tag{6.1.2}$$

$$= \frac{\omega^2}{2} a (R' M R) a , \qquad (6.1.3)$$

where K and M denote the structural stiffness and mass matrices respectively. One thus introduces the reduced matrices

$$\bar{K} = R' K R \qquad (6.1.4)$$

and

$$\bar{M} = R' M R \qquad (6.1.5)$$

defined in the new system of generalized displacements a .

The various reduction methods [4] , [23] , [26] (elimination of variables, coupling of substructures, branch modes) encountered in the literature do not essentially differ from each other, except by the choice of the deformation modes $r_{(i)}$. The influence of this choice can however be a deciding factor on the accuracy of the process, as well as on its flexibility in dealing with practical applications of various sizes.

6.2 Static condensation of variables

Consider the matrix equation governing the natural frequencies and modal shapes of the global structure in the form

$$K q = \omega^2 M q \qquad (6.2.1)$$

One possible technique, due to IRONS [27] , for reducing the size of the matrices involved consists in making a choice of a subset q_c of coordinates to be eliminated; the complementary subset being denoted by q_R (coordinates to be retained). Equation (6.2.1) is thus partitioned as follows

(6.2.2) $$K_{RR} q_R + K_{RC} q_C = \omega^2 (M_{RR} q_R + M_{RC} q_C)$$

(6.2.3) $$K_{CR} q_R + K_{CC} q_C = \omega^2 (M_{CR} q_R + M_{CC} q_C)$$

Imagine q_C to be split into two contributions

(6.2.4) $$q_C = q_S + q_D$$

where the "static" part q_S is given by

(6.2.5) $$q_S = - K_{CC}^{-1} K_{CR} q_R$$

as if in (6.2.3) one neglected the inertia forces, and a "dynamic" part q_D which is then governed by the transformed (6.2.3) equations

(6.2.6) $$(K_{CC} - \omega^2 M_{CC}) q_D = \omega^2 \bar{M}_{CR} q_R$$

with

(6.2.7) $$\bar{M}_{CR} = M_{CR} - M_{CC} K_{CC}^{-1} K_{CR}$$

The static condensation method consists in neglecting q_D and substituting directly for q_C the right-hand side of (6.2.5) into the kinetic and strain energies, which then become quadratic forms in q_R only with "reduced" stiffness and mass matrices

$$\bar{K}_{RR} = K_{RR} - K_{RC} K_{CC}^{-1} K_{CR} \tag{6.2.8}$$

$$\bar{M}_{RR} = M_{RR} - M_{RC} K_{CC}^{-1} K_{CR} - K_{RC} K_{CC}^{-1} M_{CR} + K_{RC} K_{CC}^{-1} M_{CC} K_{CC}^{-1} K_{CR} \tag{6.2.9}$$

$$= M_{RR} - M_{RC} K_{CC}^{-1} K_{CR} - K_{RC} K_{CC}^{-1} \bar{M}_{CR} \tag{6.2.10}$$

The eigenvalue problem is reduced to the condensed one

$$(\bar{K}_{RR} - \omega^2 \bar{M}_{RR})\, q_R = 0 \tag{6.2.11}$$

The procedure is of course a particular case of (6.1.4) and (6.1.5)

with

$$R = \begin{pmatrix} I \\ -K_{CC}^{-1} K_{CR} \end{pmatrix} \tag{6.2.12}$$

First order correction [2]

The validity of the condensation algorithm depends of course on the extent to which q_D is really negligible. To investigate the conditions under which this is true, consider the eigenvalue problem

(6.2.13) $$(K_{cc} - \mu^2 M_{cc})\, q_c = 0$$

to which the original problem reduces by applying the constraints

(6.2.14) $$q_R = 0$$

Let

$$\mu_1^2 \leqslant \mu_2^2 \leqslant \ldots \leqslant \mu_m^2$$

denote its eigenvalues with attendant orthonormed modal shapes

$$c_1 , c_2 \ . \ . \ . \ c_m$$

(6.2.15) $$c_1' M_{cc}\, c_j = \delta_{ij}$$

An arbitrary vector q_c as a unique expansion

$$q_c = \sum_1^m \alpha_i\, c_i$$

with coefficients determined from (6.2.15) as

$$\alpha_i = c_i' M_{cc}\, q_c$$

Thus

$$q_c = \left(\sum_1^m c_i\, c_i' M_{cc} \right) q_c$$

and, since q_c is arbitrary, we obtain the spectral expansion of the identity matrix

$$I = \sum_{1}^{m} c_i\, c_i'\, M_{cc} = \sum_{1}^{m} c_i (M_{cc}\, c_i)' \qquad (6.2.16)$$

Noting that the eigensolutions verify

$\mu_i^2 M_{cc}\, c_i = K_{cc} c_i \qquad (i = 1,2,\ldots m)$ we have also

$$I = \sum_{1}^{m} \frac{1}{\mu_i^2}\, c_i\, c_i'\, K_{cc} \qquad (6.2.17)$$

The following results derived from (6.2.16) or (6.2.17) will be of interest:

$$\omega^2 K_{cc}^{-1} M_{cc} = \sum_{1}^{m} \frac{\omega^2}{\mu_1^2}\, c_i\, c_i'\, M_{cc} \qquad (6.2.18)$$

$$K_{RC} = \sum_{1}^{m} (K_{RC}\, c_i)(M_{cc}\, c_i)' \qquad (6.2.19)$$

$$\omega^2 M_{RC} = \sum_{1}^{m} \frac{\omega^2}{\mu_i^2} (M_{RC} c_i)(K_{cc} c_i)' \qquad (6.2.20)$$

We are now able to show that the range of validity of the condensation algorithm is that for which holds

$$\frac{\omega^2}{\mu_1^2} = \varepsilon \ll 1 \qquad (6.2.21)$$

ε being the order of magnitude of the errors we are prepared to accept in the low frequency modal shapes. In other words the low frequency spectrum of (6.2.1) will be accurately represented by the eigensolution of the condensed problem (6.2.11) up to eigenvalues ω^2 satisfying (6.2.21). To show this we discuss the solutions of (6.2.6), which we put in the form

$$(6.2.22) \qquad (I - \omega^2 K_{CC}^{-1} M_{CC}) q_D = \omega^2 K_{CC}^{-1} \bar{M}_{CR} q_R$$

and (6.2.2) in which we substitute (6.2.4) and (6.2.5) to bring it into the final form

$$(6.2.23) \quad (\bar{K}_{RR} - \omega^2 \bar{M}_{RR}) q_R = (\omega^2 M_{RC} - K_{CR}) q_D + \omega^2 K_{RC} K_{CC}^{-1} \bar{M}_{CR} q_R$$

If ω^2 is such that (6.2.21) holds we can observe by comparing (6.2.18) to (6.2.16) that $\omega^2 K_{CC}^{-1} M_{CC}$ is a matrix whose elements are of order ε compared to the unit matrix. As a matter of fact (6.2.21) with $\varepsilon < 1$ is sufficient for the convergence of the expansion

$$(I - \omega^2 K_{CC}^{-1} M_{CC})^{-1} = I + \omega^2 K_{CC}^{-1} M_{CC} + (\omega^2 K_{CC}^{-1} M_{CC})^2 + \ldots$$

Comparing (6.2.19) and (6.2.20) and making the natural assumption that the vectors

$$M_{CC} c_i \qquad \text{and} \qquad M_{RC} c_i$$

are of the same order of magnitude, and similarly for

$$K_{CC} c_i \qquad \text{and} \qquad K_{RC} c_i$$

then, under (6.2.21)

$$\omega^2 M_{RC} \text{ of order } \varepsilon \text{ compared to } K_{RC} \tag{6.2.24}$$

and also

$$\omega^2 K_{CC}^{-1} M_{CC} = O(\varepsilon) \qquad \omega^2 K_{CC}^{-1} \bar{M}_{CR} = O(\varepsilon) \tag{6.2.25}$$

The zero order solution of problem (6.2.22), (6.2.23), fully equivalent to (6.2.1), is then given by

$$(\bar{K}_{RR} - \bar{\omega}^2 \bar{M}_{RR}) \bar{q}_R = 0 \qquad \bar{q}_D = 0 \tag{6.2.26}$$

where $\bar{\omega}^2$ is some eigenvalue of the condensed problem satisfying the low frequency requirement (6.2.21) and q_R the associated modal shape of the retained coordinates, while

$$\bar{q}_C = K_{CC}^{-1} K_{CR} \bar{q}_R = \bar{q}_S .$$

Equating now the terms of order ε in both equations

$$\delta q_D = \bar{\omega}^2 K_{cc}^{-1} \bar{M}_{cR} \bar{q}_R \tag{6.2.27}$$

$$(\bar{K}_{RR} - \bar{\omega}^2 \bar{M}_{RR}) \delta q_R = \delta \omega^2 \bar{M}_{RR} \bar{q}_R - K_{Rc} \delta q_D + \bar{\omega}^2 K_{Rc} K_{cc}^{-1} \bar{M}_{cR} \bar{q}_R \tag{6.2.28}$$

where δq_D and δq_R are the first order corrections for modal shape and $\delta\omega^2$ an eventual first order correction to the eigenvalue. The first order value of p_D is explicit in (6.2.27). Observing that it causes the last two terms of (6.2.28) to cancel

(6.2.29) $$(\bar{K}_{RR} - \bar{\omega}^2 \bar{M}_{RR})\,\delta q_R = \delta\omega^2 \bar{M}_{RR}\bar{q}_R$$

This is a singular system to calculate δq_R and the necessary and sufficient condition the right-hand side has to satisfy for a solution to exist is obtained by premultiplying by $\bar{q}'_R$, which will make the left-hand side vanish by virtue of (6.2.26)

$$0 = \delta\omega^2\, \bar{q}'_R\, \bar{M}_{RR}\, \bar{q}_R$$

$\bar{M}_{RR}$ being positive definite, the conclusion is that

$$\delta\,\omega^2 = 0$$

The conclusion that the eigenvalue correction is at most of second order was to be expected by virtue of the stationary character of the eigenvalues obtained from Rayleigh quotients. The only solution of (6.2.29) is now

$$\delta q_R = \delta\alpha\, \bar{q}_R$$

where $\delta\alpha$ is a, still undetermined, small scalar. There is no loss of generality in taking $\delta\alpha = 0$ since this will merely change the scale of the eigensolution, correct to second order, by a factor $(1+\delta\alpha)^{-1}$ and since

$$q_D(1 + \delta\alpha)^{-1} = \delta q_D(1 + \delta\alpha)^{-1} = \delta q_D - \delta\alpha\,\delta q_D \ldots \cong \delta q_D$$

neglecting second order terms, this will not change (6.2.27). To summarize: if the spectrum of problem (6.2.1) is to be determined accurately up to some definite frequency, the coordinates to be condensed should be numbered and chose, in such a way that the smallest eigenvalue, μ_1^2 of the constrained problem (2.6.11) sufficiently dominates the highest eigenvalue $\bar{\omega}^2$ of the condensed problem (6.2.7) that fits into the required spectrum. The ratio

$$\omega^2 / \mu_1^2 = \varepsilon \qquad (6.2.30)$$

determines the order of approximation to which the modal shapes of the condensed problem represent the exact modal shapes within the required spectrum. The first order correction on the modal shapes is given explicitly by

$$\delta q_R = 0 \qquad \delta q_C = \delta q_D \quad \text{given by (6.2.27)}$$

The eigenvalue correction is of second order.

As a last observation, if the structure has kinematical freedoms (rigid body modes or mechanism) they should not be inhibited by the choice of condensed variables.

6.3 Bound algorithms [29], [30]

In problem (6.2.1) define the successive iterates of an arbitrary starting vector q_0

(6.3.1) $$q_{p+1} = K^{-1} M q_p$$

and the corresponding Rayleigh quotients

(6.3.2) $$\rho_{2p} = \frac{q_p K q_p}{q_p' M q_p} = \frac{q_p M q_{p-1}}{q_p' M q_p} > 0$$

Further developments necessitate the definitions of the zero order Schwartz quotients

(6.3.3) $$\rho_{2p+1} = \frac{q_p' M q_p}{q_p' M q_{p+1}} > 0$$

and of the bilinear forms

(6.3.4) $$A_p(\alpha, \beta) = \frac{(q_p - \alpha q_{p+1})' K (q - \beta q_{p+1})}{q_{p+1}' K q_{p+1}}$$

(6.3.5) $$B_p(\alpha, \beta) = \frac{(q_p - \alpha q_{p+1})' M (q_p - \beta q_{p+1})}{q_{p+1}' M q_{p+1}}$$

We have

(6.3.6) $$A_p(\alpha, \beta) = \rho_{2p} \cdot \rho_{2p+1} - (\alpha + \beta)\rho_{2p+1} + \alpha\beta$$

(6.3.7) $$B_p(\alpha, \beta) = \rho_{2p+1} \cdot \rho_{2p+2} - (\alpha + \beta)\rho_{2p+2} + \alpha\beta$$

Noting that $A_p(\alpha,\alpha)$ and $B_p(\alpha,\alpha)$ are positive definite

$$\rho_{2p}\cdot\rho_{2p+1} - 2\alpha\,\rho_{2p+1} + \alpha^2 \geqslant 0 \qquad (6.3.8)$$

$$\rho_{2p+1}\cdot\rho_{2p+2} - 2\alpha\,\rho_{2p+2} + \alpha^2 \geqslant 0 \qquad (6.3.9)$$

and those expressions take their minimum value respectively for

$$\alpha = \rho_{2p+1} \quad \text{and} \quad \alpha = \rho_{2p+2}$$

For those values the inequalities (6.3.8) and (6.3.9) go into the well known Schwartz inequalities

$$\rho_{2p} \geqslant \rho_{2p+1} \geqslant \rho_{2p+2} \qquad (6.3.10)$$

Krylov-Bogoliubov bounds

Let the unique expansion of q_p in eigenmodes be

$$q_p = \sum_n a_n\, q_{(n)}$$

Then, since

$$K^{-1} M\, q_{(n)} = \frac{1}{\omega_n^2}\, q_{(n)}$$

the expansion of the next iterate will be

$$q_{p+1} = \sum_n \frac{a_n}{\omega_n^2}\, q_{(n)}$$

and, under the ortonormality relations

$$q'_{(m)} M q_{(n)} = \delta_{mn} \qquad q'_{(m)} K q_{(n)} = \omega_n^2 \delta_{mn}$$

we find

$$(6.3.11) \qquad A_p(\alpha, \beta) = \frac{\sum_n \alpha_n^2 (\omega_n^2 - \alpha)(\omega_n^2 - \beta)}{\sum_n \alpha_n^2} \quad \left(\alpha_n = \frac{a_n}{\omega_n}\right)$$

Setting $\beta = \alpha$ and comparing with (6.3.8)

$$\rho_{2p} \cdot \rho_{2p+1} - 2\alpha \rho_{2p+1} + \alpha^2 = \frac{\sum_n \alpha_n (\omega_n^2 - \alpha)^2}{\sum_n \alpha_n^2} \geqslant \min (\omega_n^2 - \alpha)^2$$

whatever be α, there is an eigenvalue ω_j^2, closest to α, and corresponding to the right-hand side minimum. Setting $\alpha = \rho_{2p+1}$, which is known to give its minimum value to the left-hand side, we find

$$\rho_{2p+1}(\rho_{2p} - \rho_{2p+1}) \geqslant (\omega_j^2 - \rho_{2p+1})^2$$

which yields the Krylov and Bogoliubov bounds

$$\rho_{2p+1} - \sqrt{\rho_{2p+1}(\rho_{2p} - \rho_{2p+1})} \leqslant \omega_j^2 \leqslant \rho_{2p+1} + \sqrt{\rho_{2p+1}(\rho_{2p} - \rho_{2p+1})}$$

(6.3.12)

In particular for any trial vector ρ_0

$$(6.3.13) \qquad \rho_1 - \sqrt{\rho_1(\rho_0 - \rho_1)} \leqslant \omega_j^2 \leqslant \rho_1 + \sqrt{\rho_1(\rho_0 - \rho_1)}$$

and if q_0 is known to be a reasonable approximation to a modal shape (6.3.13) furnishes bounds to the corresponding exact eigenvalue <u>of the discretized structure</u>. Since discretization is known to give itself upper bounds to the eigenvalues of the continuous structure, the upper bound of (6.3.13) is valid for the continuous structure. There is no similar guarantee for the lower bound.

Iterates with large values of q, based on (6.3.1), will produce close bounds (6.3.12) for the first eigenvalue ω_1^2 of the discretized structure, since the iteration process converges to this value. However, applying deflation techniques that shift already computed eigenvalues to zero, Krylov sequences of iterates can be produced whose quotients ρ_{2p} and ρ_{2p+1} converge towards the next ω_j^2. In this case (6.3.12) can produce close bounds for it. It should however be observed that in this case the upper bound is clearly worse than the simpler one

$$\omega_1^2 \leqslant \rho_{2p+1} \qquad (6.3.14)$$

Temple-Kato bounds

Let α and β lie in the interval between two consecutive eigenvalues, then it is clear from (6.3.11) that

$$A_p(\alpha,\beta) = \rho_{2p}\cdot\rho_{2p+1} - (\alpha+\beta)\rho_{2p+1} + \alpha\beta > 0 \qquad (6.3.16)$$

Take first $\alpha = \omega_i^2$

$$\omega_i^2 \leqslant \rho_{2p+1} < \beta \leqslant \omega_{i+1}^2 \tag{6.3.17}$$

Sustitution into (6.3.16) produces the bounds

$$\rho_{2p+1} \frac{\beta - \rho_{2p}}{\beta - \rho_{2p+1}} < \omega_i^2 \leqslant \rho_{2p+1} \tag{6.3.18}$$

If, for instance, ρ_{2p+1} comes from a Krylov sequence that converges to ω_i^2 the upper bound is "naturally" verified and all that is needed to obtain the lower bound to ω_i^2 is a knowledge of some lower bounds β to the next eigenvalue ω_{i+i}^2 .
The Krylov sequence can be pushed far enough for $\beta - \rho_{2p+1} > 0$ to be satisfied. If q_0 is merely an approximate modal shape associated to ω_i^2 , the inequality $\omega_i^2 \leqslant \rho_0$ is satisfied but, for $i > 1$ there is no guarantee that the iterate $q_1 = K^{-1} M q_0$ will yield $\omega_i^2 > 0$. Thus if we only work with iterates based on the original $K^{-1}M$ dynamic matrix it is preferable to consider q_0 to be the approximate modal shape in order to secure $\omega_i^2 > \rho_1$ and work backwards to compute ρ_0 from $q_0 = M^{-1} K q_1$. Even in this case it will not always be possible to know a lower bound β to ω_{i+i}^2 that verifies $\beta > \rho_1$ in order to have the lower bound

$$\rho_1 \frac{\beta - \rho_0}{\beta - \rho_1} > \omega_1^2$$

Had we worked with $B(\alpha,\beta)$ instead of $A(\alpha,\beta)$ we would have found that (6.3.18) holds also with $2p+1$ replaced by $2p+2$ and $2p$ replaced by $2p+1$. In other words if

$$\omega_i^2 \leqslant \rho_{q+1} < \beta < \omega_{i+1}^2$$
$$\rho_{q+1}\frac{\beta-\rho_q}{\beta-\rho_{q+1}} < \omega_i^2 < \rho_{q+1} \tag{6.3.19}$$

holds for any two consecutive Schwartz quotients.

Take next $\beta = \omega_i^2$ and

$$\omega_{i-1}^2 \leqslant \alpha < \rho_{2p+1} \leqslant \omega_i^2$$

Then, (6.2.16) gives

$$\rho_{2p+1} \leqslant \omega_i^2 < \rho_{2p+1}\frac{\rho_{2p}-\alpha}{\rho_{2p+1}-\alpha} \tag{6.3.20}$$

and this bounding algorithm also holds for any two consecutive Schwartz quotients. It is of a less practical value. It produces an upper bound to a given eigenvalue, from the knowledge of one of its lower bounds (ρ_{2p+1}) and of an upper bound (α) to the preceding eigenvalue. A simple way of obtaining the inequalities (6.3.19) under which (6.3.20) holds is of course to take for α a higher Schwartz quotient from a Krylov sequence converging to ω_{i-1}^2. Then, provided p and q are high enough

$$\omega_i^2 < \rho_{p+1}\frac{\rho_p-\rho_{p+q}}{\rho_{p+1}-\rho_{p+q}} \tag{6.3.21}$$

Introducing the convergence estimator

$$k_p = \sqrt{\frac{\rho_q}{\rho_{q+1}}} - 1 \tag{6.3.22}$$

of a Krylov sequence.

The Temple-Kato bounds yielded by (6.3.18) for the lower and (6.3.20) for the upper, can be put in the form

$$\rho_{q+1}\left(1 - \frac{k^2 \rho_{q+1}}{\beta - \rho_{q+1}}\right) < \omega_i^2 < \rho_{q+1}\left(1 + \frac{k^2 \rho_{q+1}}{\rho_{q+1} - \alpha}\right) \tag{6.3.23}$$

6.4 The substructure technique [16] , [21]

The concept of substructure

The method of substructures may be conceived in two different manners, as shown on Fig. 6.1.

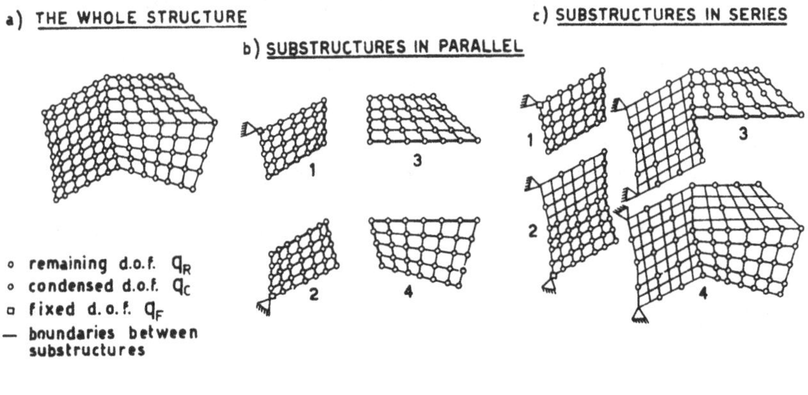

Fig. 6.1. The method of substructures

If the substructures are considered in parallel, reductions are processed separately in each of them. One then expresses transition conditions at interfaces in order to assemble the whole structure. It is required, when analizing the separate substructures, to keep the degrees of freedom in terms of which connections are expressed. The procedure is illustrated in figure 6.1a: in fact, it has to be interpreted as a method of coupling substructures that will be discussed briefly further. It may reveal successful when transition conditions between substructures involve a relatively small number of degrees of freedom.

On the other hand, substructures in series are defined in a slightly different manner, as shown in Fig. 1.c. A substructure results here from the addition to the preceding one of a certain number of finite elements. A degree of freedom can be eliminated when it is no longer required for assembling the remaining elements. One generally prefers this latter technique, since it gives a smaller maximal band with when analyzing structures that do not split naturally into strongly uncoupled subregions.

Organization of the elimination algorithm

The Nth substructure is defined as that part of the whole structure that has been assembled at the end of the preceding assembling operation. In the last step it represents the condensed form of the whole structure.

Let us denote by:

- n_e , the number of elements to be inserted in the Nth sub-structure;
- k_i and m , the element stiffness and mass matrices
- ℓ_i , the localization operators of the elements;
- $\bar{K}_{N-1}$ and $\bar{M}_{N-1}$, the reduced stiffness and mass matrices of the preceding structure;
- L_{N-1} , the localization matrix of the preceding substructure formally considered as a "super-element" for the next assembling sequence.

For each substructure the following set of operations, which is also described by the first flow chart, defines the elimination algorithm:

1) The K and M matrices of the preceding substructure are readdressed attending to their localization operator:

$$\bar{K}_N = L'_{N-1}\,\bar{K}_{N-1}\,L_{N-1} \qquad (6.4.1)$$
$$\bar{M}_N = L'_{N-1}\,\bar{M}_{N-1}\,L_{N-1}$$

2) The n_e new elements are added to the substructure:

$$K_N = \bar{K}_N + \sum_i \ell'_i\,k_i\,\ell_i \qquad (6.4.2)$$
$$M_N = \bar{M}_N + \sum_i \ell'_i\,m_i\,\ell_i$$

3) Reduction is achieved using the matrix R defined by (6.2.12):

$$\bar{K}_{RR} = R' K R$$
$$\bar{M}_{RR} = R M R \qquad (6.4.3)$$

It reduces the size of the K and M matrices, allowing space for the assemblage of the elements of the next substructure. At this stage, a displacement q can be eliminated if it satisfies the following conditions:

- it has not been specified in the list of the remaining displacements q_R
- all elements in which this displacement appears have already been included.

The last reduction operation produces the reduced matrices $\bar{K}_{RR}$ and $\bar{M}_{RR}$ of the whole structure, for which we solve the eigenvalue problem (6.2.4).

The approximate modes obtained can be restituted into the whole set of structural displacements by recalling for each substructure, in reverse order, the matrix $K_{CC}^{-1} \cdot K_{CR}$. These are stored on a disk unit during the elimination procedure,

$$q_C = - K_{CC}^{-1} K_{CR} q_R \qquad (6.4.4)$$

Error measure: computation of the first iterated vector [17] , [21]

Bounds to the error produced by the condensation are obtained by applying the algorithms (6.3.13) and (6.3.23), provided the Rayleigh and Schwartz quotients can be computed for the approximation of each eigenmode $q_{(i)}$.

First the Rayleigh quotient ρ_0 associated with the approximation q_0 of $q_{(i)}$

(6.4.5) $$q_0 = R\, q_{(i)}$$

is simply equal to the eigenvalue $\bar{\omega}_i^2$ of the reduced system:

(6.4.6) $$\rho_0 = \frac{q_0' K q_0}{q_0' M q_0} = \frac{q_{R(i)}' \bar{K}_{RR} q_{R(i)}}{q_{R(i)}' \bar{M}_{RR} q_{R(i)}} = \bar{\omega}_i^2 .$$

Next, the computation of the Schwartz quotient requires the formation of the first iterated vector, to which a slightly different definition is given from that of section 6.3:

(6.4.7) $$\rho_1 = \rho_0 K^{-1} M q_0$$

The Schwartz quotient takes thus the form

(6.4.8) $$\rho_1 = \rho_0 \frac{q_0' M q_0}{q_0' M q_1} .$$

The static problem (6.4.7) should obviously be solved without assembling again the whole structure. This is possible if, when assembling the whole structure, we store on peripheric devices all the elements needed to compute (6.4.7).

Indeed the first iterate q_1 is the solution of the linear system

$$\begin{pmatrix} K_{RR} & K_{CR} \\ K_{RC} & K_{CC} \end{pmatrix} \begin{pmatrix} q_{1R} \\ q_{1C} \end{pmatrix} = \begin{pmatrix} g_R \\ g_C \end{pmatrix}, \qquad (6.4.9)$$

where the loads g appearing in the right-hand side of (6.4.9) are the inertia loads associated with the approximate mode q_0:

$$g = \begin{pmatrix} g_R \\ g_C \end{pmatrix} = \bar{\omega}^2 \begin{pmatrix} M_{RR} & M_{RC} \\ M_{CR} & M_{CC} \end{pmatrix} \begin{pmatrix} q_{0R} \\ q_{0C} \end{pmatrix}. \qquad (6.4.10)$$

From the second equation in (6.4.9) we obtain the condensed displacements q_{1C} of the first iterated vector in terms of the q_0 :

$$q_{1C} = K_{CC}^{-1}(g_C - K_{CR}\, q_{1R}) \qquad (6.4.11)$$

By introducing equation (6.4.11) into the first equation in (6.4.9), we obtain

$$(K_{RR} - K_{RC} K_{CC}^{-1} K_{CR})\, q_{1R} = g_R - K_{RC} \cdot K_{CC}^{-1} g_C \qquad (6.4.12)$$

or

(6.4.13) $$\bar{K}_{RR}\, q_{1R} = \bar{\omega}^2 M_{RR}\, q_{0R}$$

by use of equations (6.4.10) and (6.4.).

Therefore the displacements retained take the same value for the first iterated vector q_1 and for the fundamental solution q_0 :

(6.4.14) $$q_{1R} = q_{0R}$$

this has already been pointed out in the error analysis of the elimination process. The formula (6.4.11) used to compute the condensed displacements shows that their expression is corrected by the influence of the inertia loads g_C that were neclected in the zero order approximation. In order to compute the Schwartz quotient (6.4.8), we also need the inertia loading

(6.4.15) $$g = \rho_0 M q_0$$

for the whole structure. It will be restituted by the following recurrence process: if we denote by g_R the reduced inertial load

(6.4.16) $$\bar{g}_R = \bar{\omega}^2 \bar{M}_{RR}\, q_{0R}$$

the loads g_C and g_R are given successively by

(6.4.17) $$g_C = \bar{\omega}^2 (M_{CR}\, q_{0R} + M_{CC}\, q_{0C})$$

and

$$g_R = \bar{g}_R + K_{RC} K_{CC}^{-1} g_C \tag{6.4.18}$$

To compute the first iterate (6.4.7) and the error measure coefficient

$$k^2 = \frac{\rho_0}{\rho_1} - 1 \tag{6.4.19}$$

it will thus be necessary to recall the matrices K_{CC}^{-1} , $K_{CC}^{-1} K_{CR}$, M_{CR} and M_{CC} for each substructure. This is achieved at best using direct access data set.

The second flow chart sequence gives the detailed organization of the program sequence that achieves the recursive procedure described by equations (6.4.4) and (6.4.9) to (6.4.19). We define q_0 as the eigenvector restituted in the complete set of degrees of freedom. Its components that are retained or eliminated in a specific substructure are collected in the vectors q_{OR} and q_{OR} respectively; the same notations hold for the first iterate q_1 and the inertia load g . The transfer operations from q_0 to q_{OR} and q_{OC} (denoted by an arrow in the flow chart) are performed by recalling the localization vector of the substructure: it gives the address of each component of q_{OR} or q_{OC} in q_0.

Finally it is important to note that the error measure coefficient, k^2 , will only be used to produce lower bounds to the exact eigenfrequencies; indeed we have establish ed in section 6.2 that the computed eigenfrequencies $\bar{\omega}^2$ are the most accurate upper bounds to the eigenvalues of equation (2.1.1).

6.5 The coupling methods

When solving approximately the eigenvalue problem of large structures by synthetising the results of a parallel analysis of the constitutive parts, the difficulty lies in the choice of the deformation modes to be adopted for representing the motion of the different subregions.
Indeed these modes must satisfy two essential conditions:

- to give an accurate representation of the internal behaviour for each subregion;
- to allow an easy expression of transition conditions along interfaces.

When using the method of substructures in parallel, the generalized displacements of each substructure are split in two categories:

- the boundary degrees of freedom, q_B
- the interior degrees of freedom, q_I.

By analogy with the assembling process of kinematically admissible elements, an accurate representation of the static behaviour of a substructure interacting with the adjacent ones implies that the conformity of the displacement field along the interface is preserved. In other words, we have to include all deformation modes generated by any sollicitation of interfaces. They are collected in the modal matrix

$$R_L = \begin{pmatrix} I \\ \Phi_L \end{pmatrix} = \begin{pmatrix} I \\ -K_{II}^{-1} \cdot K_{IL} \end{pmatrix} . \tag{6.5.1}$$

We superimpose to the boundary modes (6.5.1) a certain number of interior deformation modes selected as the most representative of the substructure considered. They may be restricted without loss of generality to the form

$$R_I = \begin{pmatrix} 0 \\ \Phi_I \end{pmatrix} \tag{6.5.2}$$

which is obtained by imposing zero boundary displacements.

Hurty's method [26]

When the interior deformation modes Φ_I chosen are part of the vibration modes Φ_N of the substructure, one obtains the method of substructures coupling proposed by HURTY [26] and used in conjuction with the finite element method by CRAIG and BAMPTON [5] .

The normal modes Φ_N are solutions of the interior eigenvalue problem

$$K_{II} \Phi_N = \omega^2 M_{II} \Phi_N , \tag{6.5.3}$$

and the reduction matrix takes the form

$$R = \begin{pmatrix} I & 0 \\ \Phi_C & \Phi_N \end{pmatrix} .$$

One obtains for each substructure the reduced stiffness and mass matrices

$$(6.5.4) \qquad \bar{K} = \begin{pmatrix} \bar{K}_{LL} & 0 \\ 0 & \bar{K}_{NN} \end{pmatrix} \qquad \bar{M} = \begin{pmatrix} \bar{M}_{LL} & 0 \\ 0 & \bar{M}_{NN} \end{pmatrix}$$

with the submatrices

$$\bar{K}_{LL} = K_{LL} - K_{LI} K_{II}^{-1} K_{IL}$$

$$\bar{K}_{NN} = \Phi_N' K_{II} \Phi_N$$

$$(6.5.5) \qquad \bar{M}_{LL} = M_{LL} + \Phi_L' M_{II} \Phi_L + M_{LI} \Phi_L + (M_{LI} \Phi_L)'$$

$$\bar{M}_{LN} = \bar{M}_{NL}' = \Phi_L' M_{II} \Phi_N + M_{LI} \Phi_N$$

$$\bar{M}_{NN} = \Phi_N' M_{II} \Phi_N$$

Hurty's method seems particularly well adapted to the eigenvalue computation of, for instance, a complete aircraft structure as a combination of the eigenmodes of each part analyzed independently. Indeed, the eigenmodes of each part of the whole structure are strongly uncoupled from each other, as it is generally observed experimentally from vibration tests. This is due to the fact that the eigenfrequencies of the fuselage are very different from those of the lifting surfaces.

The use of static modes

The vibration modes used in Hurty's method are specific of the high frequency behaviour of the substructure. It can therefore be objected that the method does not lead to a correct representation of the quasi-static behaviour of a substructure considered as a part of the whole structure. Refining the subdivision into substructures reinforces this weakness.

Hence another way of reducing the internal discretization of the substructure consists to admit, as a first approximation, that the internal deformation modes selected are those obtained by static application of the inertia loads generated under rigid body motion.

In other words, the set of normal modes proposed by Hurty are replaced by the set of static modes obtained by application to the substructure of the inertia forces conjugated to its rigid body modes.

This latter procedure also presents the advantage of avoiding the solution of the eigenvalue problem for each substructure successively, and replacing it by a static analysis, the cost of which is much reduced.

Its application to plate structures [22] has shown that an important reduction of the number of degrees of freedom can be achieved without any significant alteration of the lower frequency spectrum.

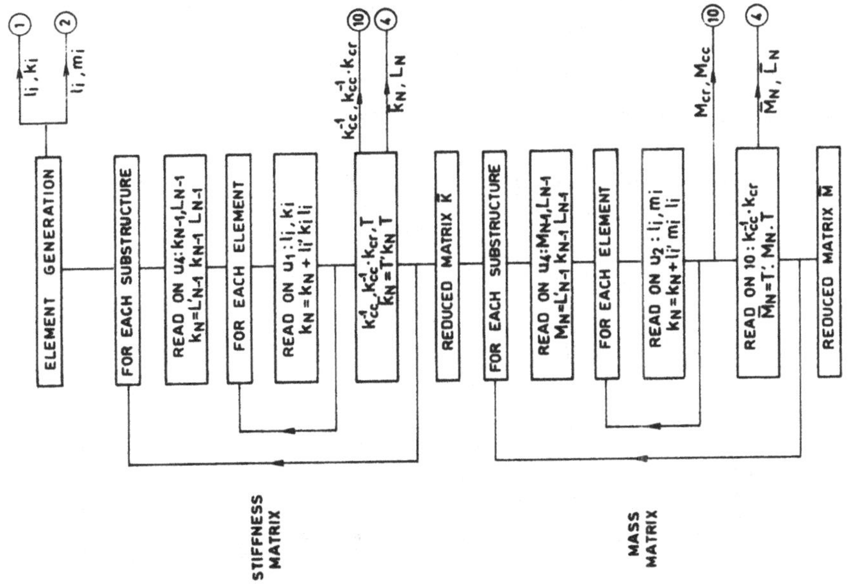

Fig. 6.2. Flow chart NR 1: The assembling sequence

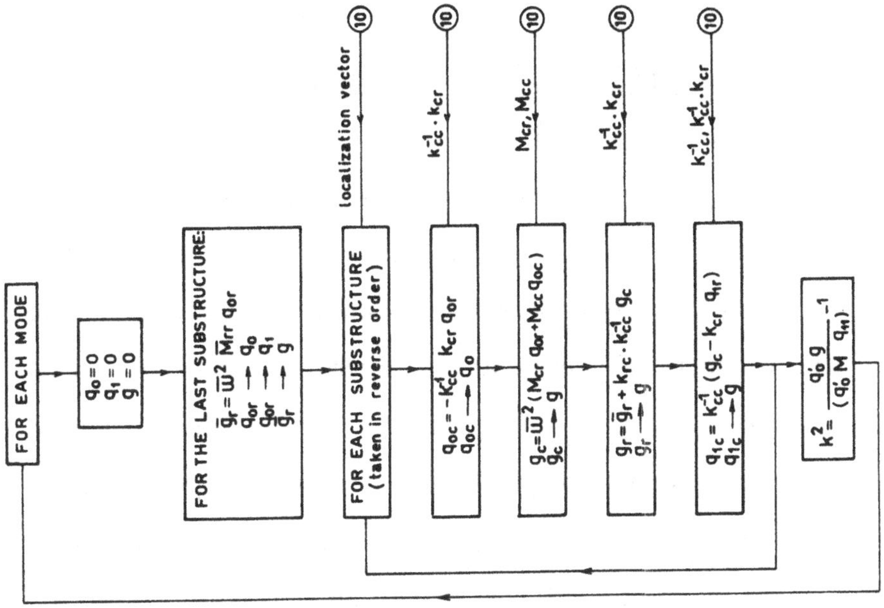

Fig. 6.3. Flow chart NR 2: The restitution sequence

7. NUMERICAL APPLICATIONS OF THE DUAL ANALYSIS TO PLATE LIKE STRUCTURES

The numerical applications presented are limited to the dual analysis of thin plate structures. The corresponding plate bending finite elements are illustrated on figure 7.1. The conforming displacement elements are derived as shown by figure 7.2 by assembling triangular subregions to form a super element [9] , [33] . In each subregion a cubic polynomial represents the deflection. The degrees of freedom are the deflection and two slopes at each vertex, and the slopes normal to the external interfaces.

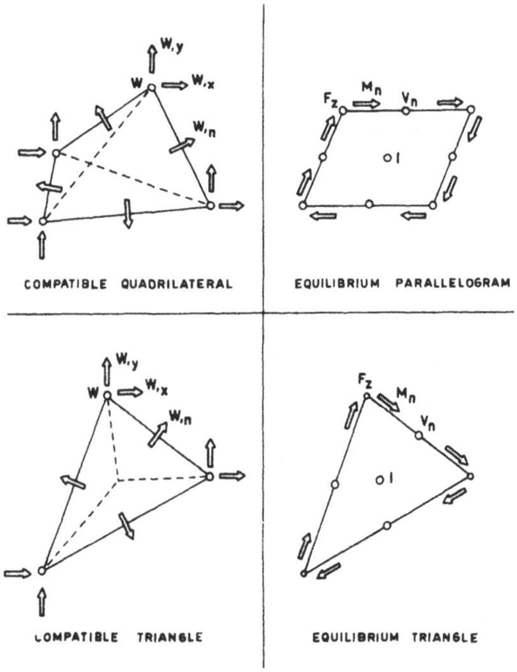

Fig. 7.1. Plate bending finite elements for dual analysis.

In the equilibrium models, parallelogram and triangle shapes are used. A linear variation of the bending moment is assumed, as shown by the figure 7.3. The

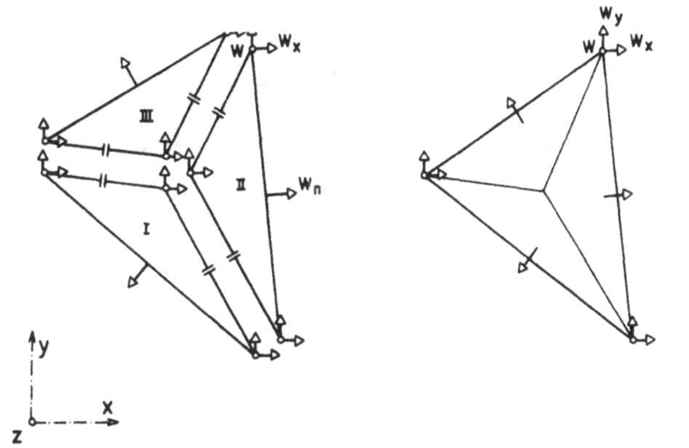

Fig. 7.2. Kinematical plate element as an assembling of three triangular regions

variables are concentrated loads at vertices, bending moments and Kirchhoff shear load along interfaces, plus the internal degrees of freedom corresponding to the inertia forces which correspond to the three rigid body modes of the element.

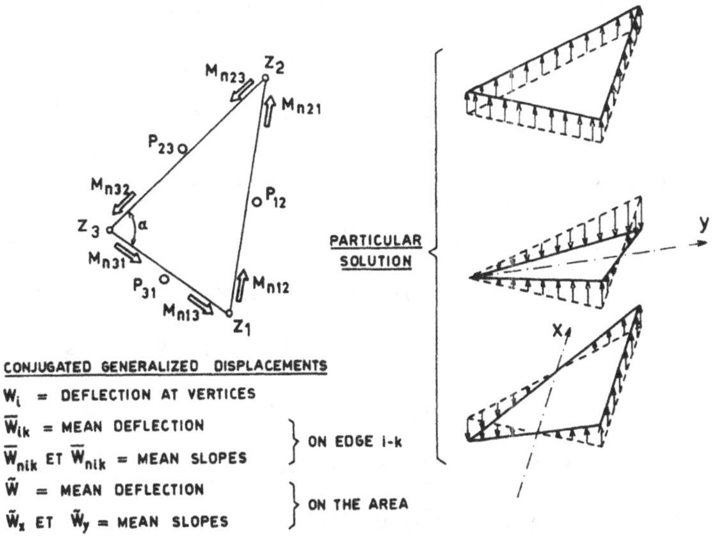

Fig. 7.3. Equilibrium plate element

7.1 Cantilever square plate

Figure 7.4 illustrate the application of the dual analysis principle to a cantilever plate of uniform thickness. It has been analyzed first with only one element for the whole plate, and then using finer and finer grid up to 36 elements which corresponds roughly to 200 degrees of freedom in the kinematical approach conforming quadrilateral element denoted CQ . Two subdivisions of the elementary square mesh into equilibrium triangular elements (EQT) have been adopted for the dual approach. The finest idealization - note that it corresponds to the 3×3 subdivision - involves 600 degrees of freedom, from which 216 only contribute to the kinetic energy, and hence are retained in the eigenvalue problem.

One can appreciate the monotonic convergence properties of the finite element solution, and the effectiveness and usefullness of the bounds computed to the eigenfrequencies. With relatively few elements, good approximations of the five first eigenfrequencies are obtained, and the gap betwen bounds has been reduced so that their average value represents for practical purposes an exact solution (difference less than 0,2% for the five computed eigenfrequencies). It is also worth pointing that equilibrium models yield for crude idealization better results than compatible elements.

non dimensional circular frequency

$$\bar{\omega} = \frac{\omega a^2}{\sqrt{\frac{D}{m}}}$$

m = mass/unit of area

D = bending rigidity

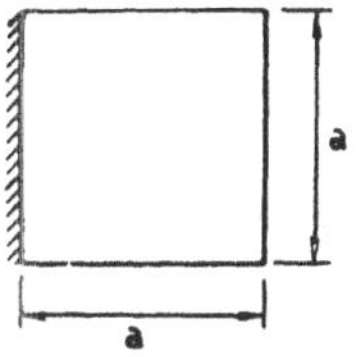

DISPLACEMENT APPROACH

MODE N°	1 X 1	2 X 2	3 X 3	4 X 4	5 X 5	6 X 6
1	3.489	3.486	3.479	3.476	3.474	3.473
2	9.222	8.606	8.535	8.518	8.513	8.511
3	26.96	21.50	21.38	21.33	21.31	21.30
4	35.29	27.63	27.39	27.28	27.24	27.22
5	43.75	31.97	31.27	31.08	31.02	30.99

EQUILIBRIUM APPROACH

2 SUBDIVISIONS:

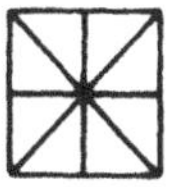

	1 X 1		2 X 2		3 X 3		4 X 4
MODE N°	⊠	⧆	⊠	⧆	⊠	⧆	⊠
1	3.447	3.453	3.462	3.466	3.467	3.469	3.468
2	7.850	8.246	8.410	8.470	8.475	8.494	8.492
3	20.74	20.76	21.16	21.21	21.24	21.26	21.26
4	25.07	26.73	26.89	27.07	27.10	27.16	27.16
5	28.86	28.92	30.14	30.72	30.75	30.89	30.87

Fig. 7.4. Square cantilever plate

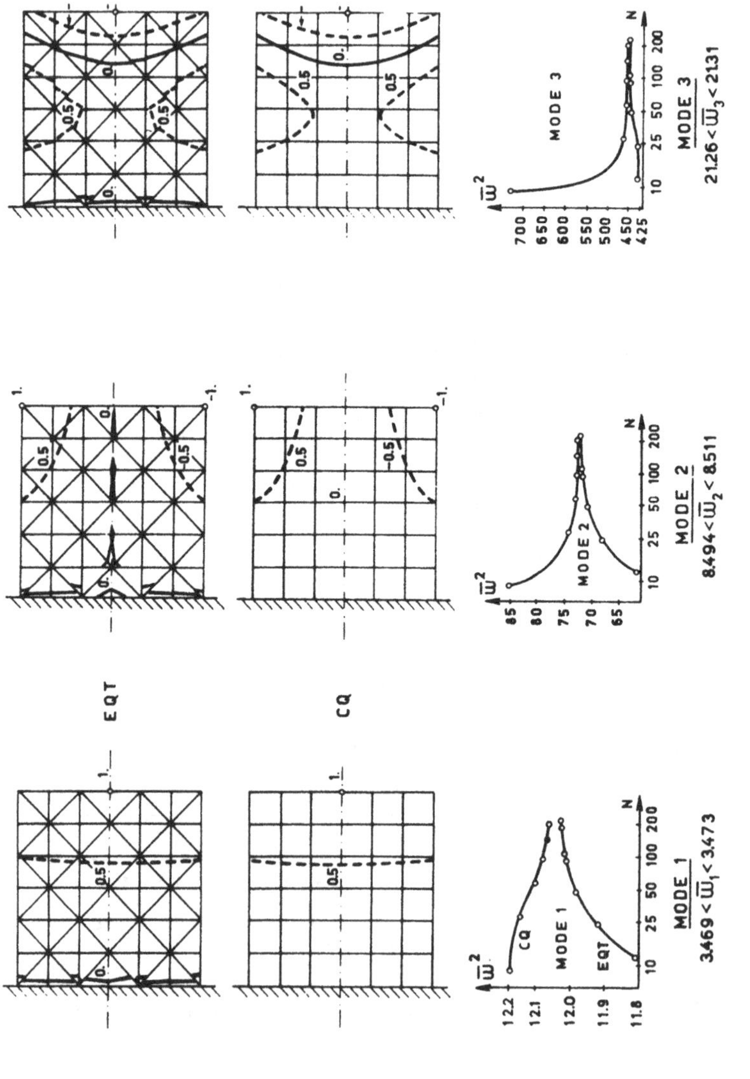

Fig. 7.5. Square cantilever plate

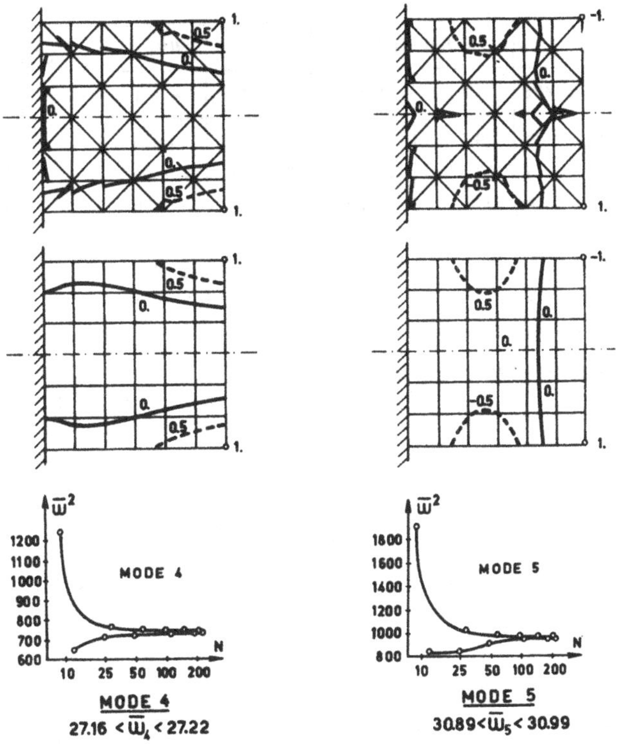

Fig. 7.6. Square cantilever plate

Figures 7.5 and 7.6 represent the eigenmodes obtained from both approaches, as well as the convergence associated with both types of models when increasing the number of degrees of freedom. The modes obtained by using equilibrium models correspond to a "weak knowledge" of the displacement field. It is thus necessary to compute (from the knowledge of the generalized displacements conjugated to the inertia forces only) an average displacement of the element as a rigid body. The discontinuities of the nodal lines translate the mathematical concept of dislocation which is specific of equilibrium models.

7.2 Point supported plates

As the point support represents for plate bend-

		EQUILIBRIUM		CONFORMING	
MODE	TYPE	4 x 4	8 x 8	12 x 12	4 x 4
1	S-S	19.594	19.596	19.596	19.616
2	S-S	23.335	23.368	23.380	23.426
3	S-A	31.540	32.323	32.642	33.175
4	A-A	33.658	34.659	35.074	35.489
5	S-A	35.087	35.214	35.283	35.917
6	S-S	—	54.077	55.365	59.714

NON DIMENSIONAL FREQUENCY $\frac{\omega a^2}{\sqrt{D/m}}$

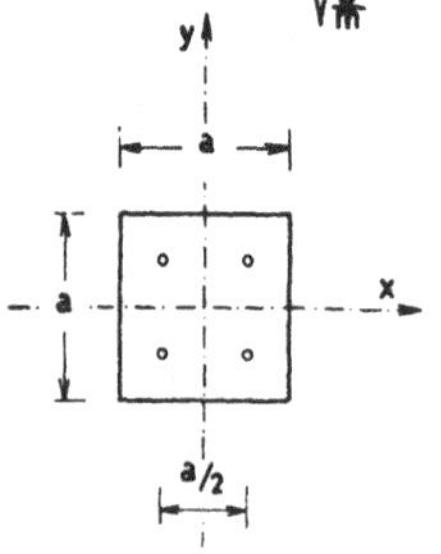

Fig. 7.7. Bounds to eigenfrequencies by dual analysis point supported square plate.

ing problem a severe test due to the singularity at these bending points, the accuracy obtained by finite elements has therefore been tested on a square plate point supported in the center of each quarter plate (figures 7.7 and 7.8). Two grids of finite elements have been used: 4 x 4 and 8 x 8 in the case of e-

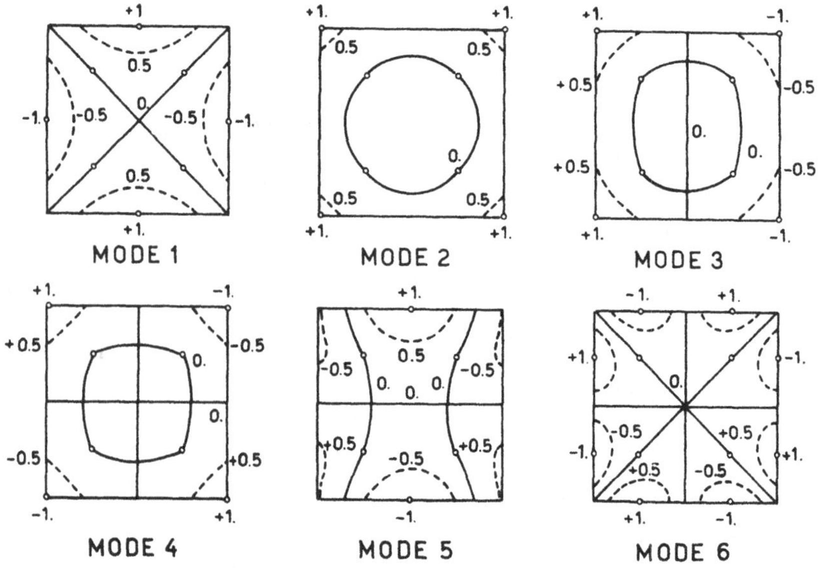

Fig. 7.8. Vibration modes of a point supported square plate.

quilibrium models, 4×4 and 12×12 in the case of conforming models. The finer CQ subdivision involves 541 degrees of freedom, and account was taken of the symmetry to solve the eigenvalue problem. The same remark holds for the second equilibrium approach, since it counts 2133 generalized displacements, from which 768 contribute to the kinetic energy.

The very good agreement between the dual analysis, which yield very narrow brackets for the exact eigenvalue solution, allows to be very confident in the behavior of these elements in the case of point support.

7.3 Cantilever skew plates

Figures 7.9 to 7.14 illustrate the dual solution of a cantilever plate for various skew angles.

In the displacement analysis by CQ elements, the finest grid corresponds as before to 231 degrees of freedom, from which 27 are fixed. The results collected in the figure 7.9 confirm the upper boundeness guaranteed by the kinematical approach. Note that the finite element solution deteriorates for increasing skewness.

The equilibrium approach (figure 7.10) underlines the same phenomenon. The finest mesh uses now 50 elements and 420 free generalized displacements. Only 150 of them appear in the eigenvalue problem. Note that in some cases the computed

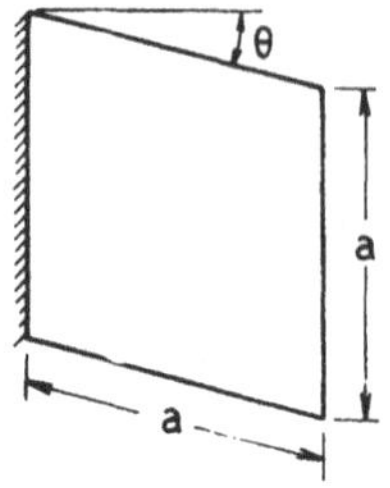

non dimensional circular frequency

$$\bar{\omega} = \frac{\omega a^2}{\sqrt{\frac{D}{m}}}$$

m = mass/unit of area

D = bending rigidity

DISPLACEMENT APPROACH

	1x1	2x2	3x3	4x4	5x5	6x6
	3.652	3.602	3.591	3.587	3.586	3.585
	9.376	3.872	8.765	8.731	8.717	8.710
15°	27.88	22.66	22.42	22.32	22.28	22.26
	35.21	26.78	26.52	26.41	26.37	26.36
	47.42	35.14	34.38	34.10	33.99	33.94
	44.159	3.988	3.954	3.943	3.938	3.935
	10.12	9.683	9.532	9.480	9.456	9.443
30°	28.72	26.22	25.75	25.56	25.47	25.41
	40.09	27.08	26.36	26.12	26.03	25.99
	57.33	43.03	42.06	41.67	41.52	41.45
	5.070	4.722	4.624	4.532	4.560	4.547
	12.19	11.47	11.37	11.32	11.30	11.29
45°	31.42	28.43	27.65	27.37	27.25	27.18
	48.37	34.15	32.62	32.18	31.98	31.86
	72.04	56.91	52.35	51.38	51.04	50.90

Fig. 7.9. Skew cantilever plates.

frequency appears now to be an upper estimation. Hence we conclude that the lower bound properties expected from the equilibrium models can fail when the discretization adopted reveals too crude for a correct representation of inertia forces.

An other important factor is the type of subdivision adopted for triangular elements: its influence increases with the skewness of the plate.

All these remarks strike up from the examination of the figure 7.11 and 7.12 which represent, for modes 1 and 2, the convergence as a function of three factors:

- The number of degrees of freedom in terms of which the eigenvalue problem is solved;
- the skewness of the plate;
- for triangular equilibrium models, the type of subdivision

EQUILIBRIUM APPROACH

3 TYPES OF SUBDIVISION

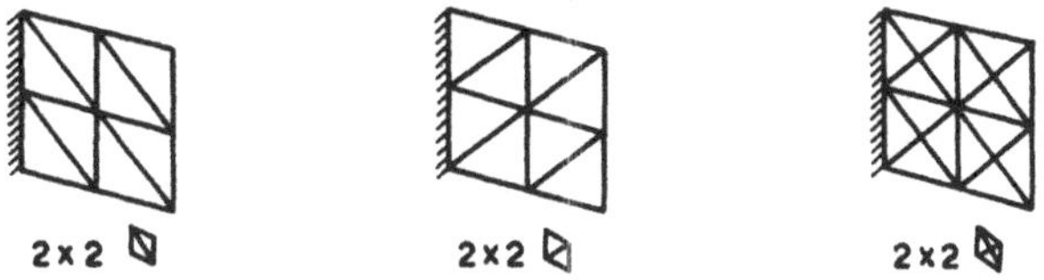

	1 X 1			2 X 2			3 X 3			4 X 4		5 X 5
15°	3.378 7.374 11.82 22.35 39.37	3.400 7.733 16.02 20.18 47.97	3.502 8.096 21.46 24.03 32.46	3.537 8.377 21.23 25.60 31.17	3.549 8.478 21.64 25.58 32.54	3.567 8.583 21.97 25.93 33.03	3.567 8.571 21.84 25.93 32.89	3.565 8.607 22.00 26.12 33.32	3.570 8.651 22.13 26.24 33.59	3.576 8.632 22.05 26.13 33.41	3.574 8.649 22.11 26.25 33.60	3.579 8.667 22.16 26.29 33.71
30°	3.374 7.917 10.95 23.11 37.92	3.562 8.681 19.17 21.08 45.15	3.709 8.884 22.89 25.08 36.79	3.758 8.929 22.07 25.10 34.88	3.348 9.169 24.05 25.33 40.02	3.830 9.257 24.56 25.53 40.33	3.857 9.166 23.91 25.32 39.14	3.892 9.297 24.76 25.63 40.82	3.906 9.338 24.97 25.73 41.03	3.689 9.266 24.57 25.60 40.34	3.907 9.338 24.97 25.78 41.03	3.914 9.365 25.09 25.85 41.18
45°	3.344 8.302 11.66 22.30 40.31	3.942 10.26 23.35 26.83 43.01	4.084 10.43 26.27 31.56 40.71	4.002 10.34 22.69 25.83 35.32	4.321 11.00 25.57 30.37 48.09	4.370 11.06 26.01 30.40 48.79	4.235 10.62 24.67 28.54 46.37	4.406 11.14 26.35 30.84 49.97	4.432 11.16 26.53 30.98 50.27	4.335 11.00 25.56 29.87 48.16	4.441 11.13 26.60 31.06 30.41	4.460 11.20 26.72 31.17 50.56

Fig. 7.10. Skew cantilever plates

adopted.

In order to facilitate the interpretation of the diagrams, the curves associated to different skew angles have been shifted horizontally.

The deterioration of the eigenvalue solution with the skewness of the plate becomes evident. Note also the considerable influence of the type of subdivision adopted.

The figure 7.13 represents the modes 1 and 2 for $\theta = 45°$; those corresponding to equilibrium models have to be interpreted in the same way as described for the square plate (section 6.1).

Finally the results reported on figure 7.14 allow a comparison with analyses realized by other authors. Among

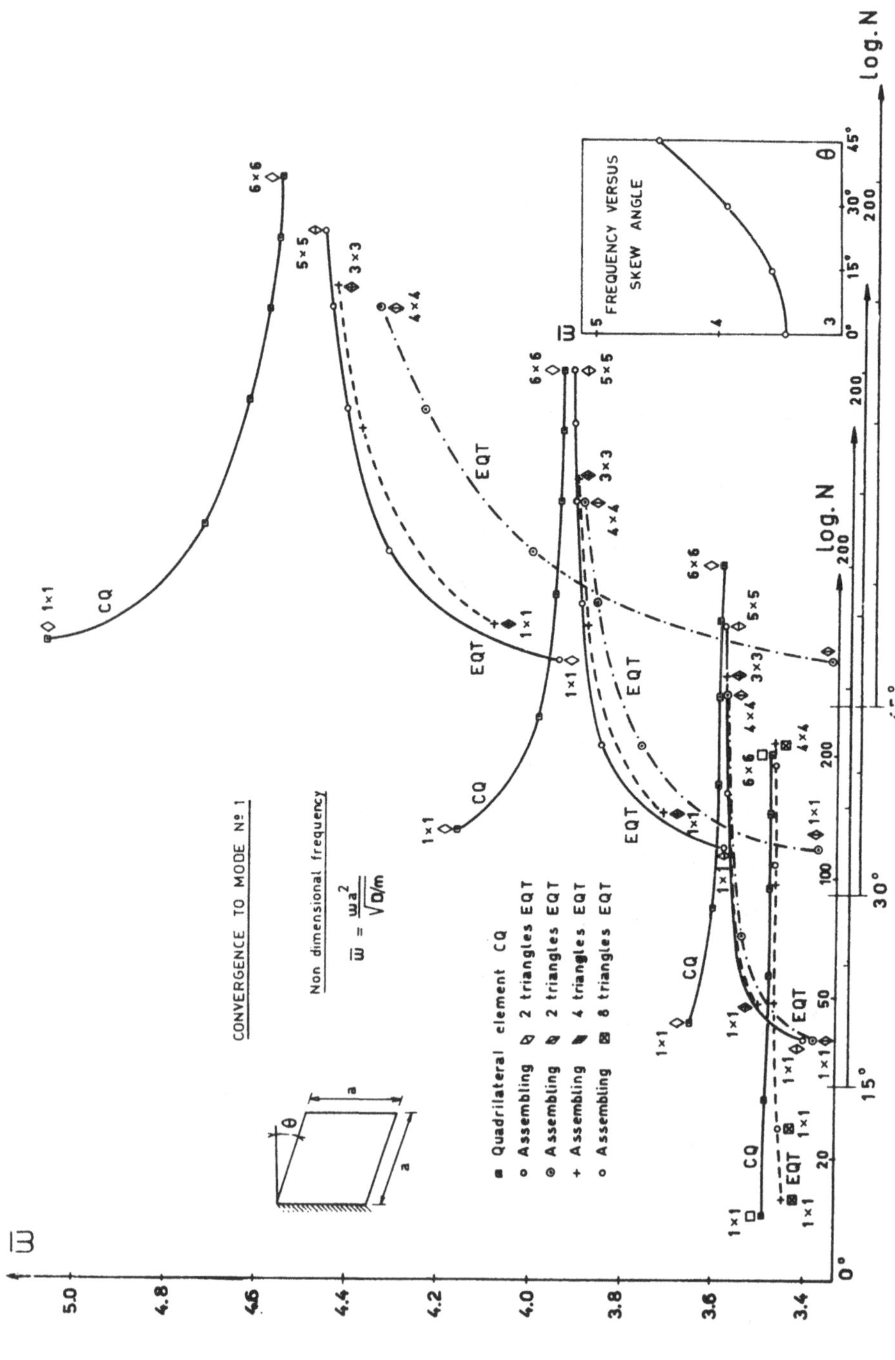

Fig. 7.11. Dual analysis of skew cantilever plates

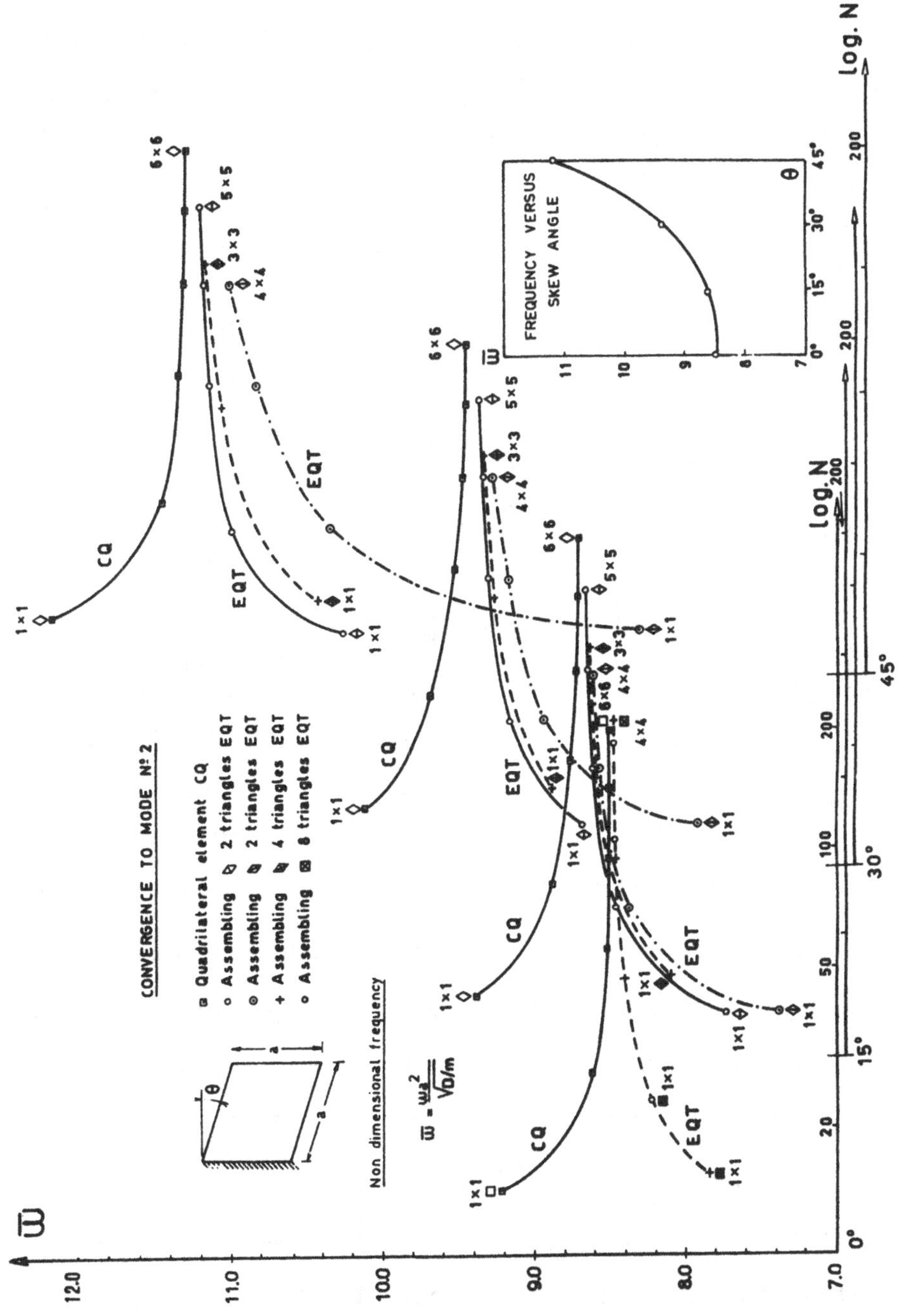

Fig. 7.12. Dual analysis of skew cantilever plates

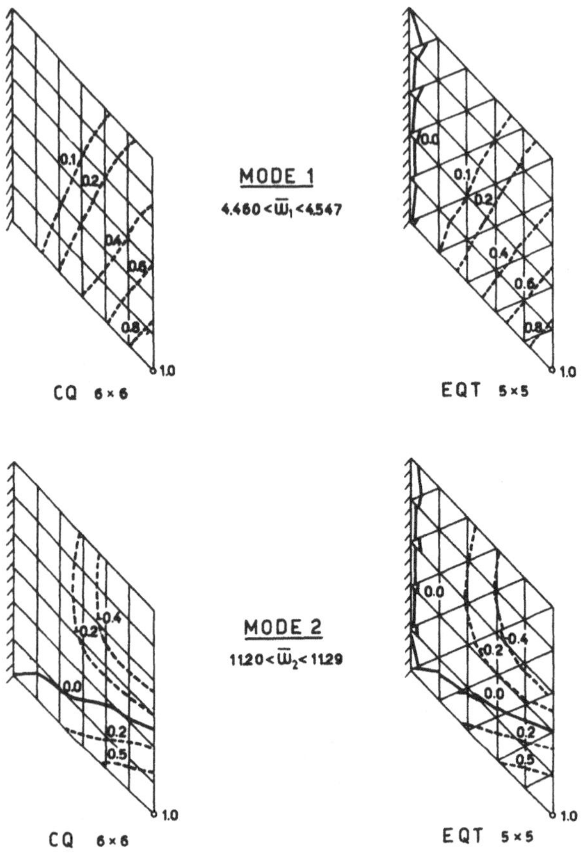

Fig. 7.13. Skew cantilever plate 45°.

others, note that the experimental eigenfrequencies given by Barton [1] , even after the correction compensating the additional inertia due to aerodynamic forces, are not bracketed by the bounds obtained from the dual analysis. The advantage offered by a dual analysis appears clearly in this case, as it allows a better understanding of the discrepancies between experimental results and an analysis. The difference between experiment and numerical approach is very likely due to the difficulty to achieve a perfectly clamped support.

The comparison with other numerical results also underlines that non conforming kinematical elements [43] do not guarantee upper bounds to exact eigenfrequencies.

	THEORETICAL (1)	TEST (1)	CORRECTED TEST (1)	DAWE (2)	ZIENKIEWICZ (3)	CQ 6×6 (4)	EQT 5×5 (4)	ΔωZ (4)
15°	3.60	3.38	3.44	3.59	3.57	3.53	3.58	0,2
	8.87	8.63	8.68	8.71	8.60	8.71	8.67	0,5
	-	21.49	-	21.59	21.75	22.26	22.16	0,5
	-	26.04	-	-	-	26.36	26.23	0,3
	-	33.01	-	-	-	33.94	33.71	0,7
30°	3.96	3.82	3.38	3.95	3.98	3.93	3.91	0,5
	10.19	9.23	9.33	9.42	9.19	9.44	9.37	0,8
	-	24.51	-	25.56	24.56	25.41	25.09	1,3
	-	25.54	-	-	-	25.99	25.85	0,6
	-	40.64	-	-	-	41.45	41.18	0,7
45°	4.82	4.26	4.33	4.59	4.67	4.55	4.46	2,0
	13.75	11.07	11.21	11.14	11.01	11.29	11.20	0,8
	-	26.52	-	27.47	27.56	27.18	26.72	1,7
	-	30.13	-	-	-	31.36	31.17	2,2
	-	50.19	-	-	-	59.90	56.56	0,7

(1) Theory (RAYLEIGH-RITZ) and experiment
BARTON, Jnl App. Mech., 1951.

(2) Parallelogram element (4 × 4, 75 d.o.f.):
DAWE, Jnl of Strain Analysis, 1966.

(3) Triangular element (8 elements, 18 d.o.f.):
ZIENKIEWICZ, Int. Jnl of Solids and Structures, 1968.

(4) Displacement and equilibrium approaches with CQ and EQT elements.

Fig. 7.14. Skew cantilever plates camparison of results

REFERENCES

[1] Barton, M.V., 1951, Jnl. Appl. Mech. 18, 129-134 "Vibration of rectangular and skew cantilever plates".

[2] Courant, R. and Hilbert, D., 1937, Interscience Publishers, N.Y., "Methods of Mathematical Physics", vol. 1, (English Edition, 1953).

[3] Cox, H.L., 1971, Aircraft Engineering, vol. XXXIII, n° 383, pp 2-7, "Vibration of missiles", part 1.

[4] Craig, R. and Bampton M., 1968, AIAA Jnl. vol. 6, n° 7, pp. 1313-1319, "Coupling of substructures for dynamic analysis".

[5] Craig, R. and Bampton, M., 1971, The Aeronautical Jnl. of the R.A.S., vol. 75, pp. 287-290, "On the iterative solution of semi-definite eigenvalue problems".

[6] Crandall, S.H., 1957, Proc. 9th int. Cong. Appl. Mech. (Brussels), vol. 5, pp. 80-87, "Complementary extremum principle for dynamics".

[7] Dawe, D.J., 1966, Jnl. Strain Analysis, vol. 1, n° 3, "Parallelogrammic elements in the solution of rhombic cantilever plate problems".

[8] Fraeijs de Veubeke, B.M., 1955, Jnl of Aeron. Sciences, vol. 22, n° 10, pp. 710-720, "Iteration in semi definite eigenvalue problems".

[9] Fraeijs de Veubeke, B.M., 1968, Int. Jnl. of Solids and Structures, vol. 4, "A conforming element for plate bending".

[10] Fraeijs de Veubeke, B.M. and Sander, G., 1968, Int. Jnl.

of Solids and Structures, vol. 4, "An equilibrium element for plate bending".

[11] Fraeijs de Veubeke, B.M., Annales de la Société Scientifique de Bruxelles, T. 70, n° 1, pp. 37-61, 1956, "Matrices de projection et techniques d'itération".

[12] Fraeijs de Veubeke, B.M., 1971, Nato Advanced Study Institute on finite element methods in continuum mechanics (LISBON), "Duality in structural analysis by finite elements".

[13] Fraeijs de Veubeke, B.M., 1971, Nato Advanced Study Institute on finite element methods in continuum mechanics (LISBON), "The dual principle of elastodynamics - finite element applications". The University of Alabama in Huntsville Press 1972.

[14] Fraeijs de Veubeke, B.M., 1965, Fluid Dynamics Transactions, Proceedings of the Polish Conference of Jurata, Warsaw, "Variational Principle in Fluid Mechanics."

[15] Fraeijs de Veubeke, B.M., 1965, ch. 9 of Stress Analysis, Wiley, "Displacement and equilibrium models in the finite element method".

[16] Geradin, M., 1970, LTAS, Université de Liège, Rpt Vf-9 "Notice d'utilization du programme d'analyse dynamique DYNAMG".

[17] Geradin, M., 1971, Jnl of Sound and Vibration, vol. 19, n° 1, "Error bounds for eigenvalues analysis by elimination of variables".

[18] Geradin, M., Jnl of Sound and Vibration, vol. 19, N°3 "The computational efficiency of a new minimization algorithm for eigenvalue analysis"

[19] Geradin , M., 1971, Symposium IUTAM "High Speed Computing of Elastic Structures, Congrès et Colloques de l'Université de Liège, Place du XX Août, 16, B-4000 LIEGE.
"Computational efficiency of equilibrium models in eigenvalue analysis".

[20] Geradin , M., 1971, LTAS, Université de Liège, Rpt VA-5 "The assumed stress variational principles in elastodynamics".

[21] Geradin , M., Faculté des Sciences Appliquées de Liège, Ph.D. Thesis, 1972
"Analyse dynamique duale des structures par la méthode des éléments finis".

[22] Geradin , M., 1969, LTAS, Université de Liège, Rpt VF-5 "Analyse dynamique de structures complexes par combinaison linéaire de modes statiques".

[23] Gladwell, G.M.L., 1964, Jnl of Sound and Vibration, vol. 1 41-59, "Branch modes analysis of vibrating systems".

[24] Gladwell, G.M.L. and Zimmermann, G., 1966, Jnl of Sound and Vibration, vol. 3, 233-241, " On energy and complementary formulations of acoustic and structural vibration problems".

[25] Gould, S.H., 1957, University of Toronto, "Variational methods for eigenvalue problems".

[26] Hurty, 1965, AIAA Jnl, vol. 3, n°4, 678-685, "Dynamic analysis of structural systems using component modes".

[27] Irons, B.M., 1965, AIAA Jnl, vol.3, n°5, 961-962, "Structural eigenvalue problems: elimination of unwanted variables".

[28] Kato, T., 1949, J. Phys. Soc. Japan, vol.4, 334, "On the upper and lower bounds of Eigenvalues".

[29] Kestens, J., 1956, Mémoires de l'Académie Royale de Belgique, Classe des Sciences, vol. 29, n° 4, "Le problème aux valeurs propres normal et bornes supérieures et infèrieures par la méthode des itérations".

[30] Kestens, J., 1961, Mémoires de l'Académie Royale de Belgique, Classe des Sciences, vol. 32, n° 5, Le problème naturel aux valeurs propres".

[31] Penrose, R., 1951, Proc. Cambridge Phil. Soc., vol. 51, pp. 406-413, "A generalized inverse for matrices".

[32] Sander, G., 1969, Collection des Publications de la Faculté des Sciences Appliquées, Liège, Ph.D. Thesis, "Application de la méthode des éléments finis à la flexion des plaques".

[33] Sander, G., 1970, Symposiom IUTAM, "High Speed Computing of elastic Structures", Congrès et Colloques de l'Université de Liège, Place du XX Août , 16, B-4000 Liège, "Application of the dual analysis principle".

[34] Sander, G. and Beckers, P., 1971, 4th Conference on Matrix Methods in Structural Mechanics Wright Patterson AFB, OHIO, "Improvements of finite element solutions for structural and non structural applications".

[35] Tabarrok, B. and Karnopp, B.H., 1968, Jnl of Sound and Vibration, vol. 8, n°3, pp. 469-481, "On duality in the oscillations of framed structures".

[36] Tabarrok, B. and Sakaguchi, R.L., 1970, Int. Jnl for Num. Meth. in Eng. vol. 2, 283-293, "Calculations of plate frequencies from complementary energy formulation".

[37] Temple, G., 1952, Proc. Roy. Soc. A211, pp. 204-224,

"The accuracy of Rayleigh's method of calculating the natural frequencies of vibrating systems".

[38] Toupin, R.A., 1952, Jnl Appl. Mech., "A variational principle for the mesh-type analysis of a mechanical system".

[39] Wilkinson, J.H., 1965, Clarendon Press, Oxford, "The algebraic eigenvalue problem".

[40] Wilkinson, J.H., 1961, Ch. 3 of: "Mathematical methods for digital computers," vol. 2, Ed. Ralston and Wilf, Wiley; "The solution of ill-conditioned linear equations".

[41] Yu Chen, 1964, Jnl Franklin Inst., vol. 278, n°1, pp. 1-7, "Remarks on variational principles in elastodynamics".

[42] Zienkiewicz, O.C., 1971, McGraw Hill, London; "The finite Element method in engineering science".

[43] Zienkiewicz, O.C., Irons, B.M. and Anderson, R.G., 1968, Jnl of Solids and Structures, 4, 1031-1035, "Vibration and stability of plates using finite elements".

8. TRANSIENT RESPONSE METHODS BASED ON MODAL ANALYSIS

8.I Introduction

When a linear elastic structure has been idealized into a discrete undamped system, by means of the finite element method or any other discretization procedure, the basic equation governing its motion under external time dependent forces is:

$$M\ddot{q} + Kq = f(t) \tag{8.1}$$

where M and K are the symmetric mass and stiffness matrices, respectively; q and $\ddot{q}$ are the generalized displacements and acceleration vectors, respectively, and $f(t)$ is the generalized force vector.

The mass matrix M is positive definite while the stiffness matrix K may be only non negative if the system has internal mechanism or rigid body degrees of freedom.

In order to solve the basic equation (8.1), two major classes of methods can be used:
The first alternative consists in the direct integration of the coupled equations of motion (8.1) by means of finite differences or any other numerical integration procedure.
The other one is based on a preliminary modal analysis of the

system leading to the uncoupled equations of motion which may the be integrated separately in order to get the general solution of equation (8.1) by superposition of the displacements in each eigenmode shape.

In spite of strong difficulties relative to stability and convergence, especially when response is to be computed over a large time range, the direct method seems to be often preferred to the modal decomposition one. It avoids a rather extensive modal analysis that may reveal prohibitive, for large systems, mainly when high frequency excitation is significant.

The purpose of the present paper is to show that such a choice is not always justified and that methods, based on a partial modal analysis including a sufficient number of significant low frequency modes, can be used successfully in many transient response problems.

8.2 Modal analysis of linear elastic systems

Free vibrations of a discrete linear elastic system are governed by the homogeneous form of basic equation (8.1):

$$M\ddot{q} + Kq = 0 \tag{8.2}$$

which has particular solutions of the form:

$$q = x e^{i(\omega t + \phi)} \tag{8.3}$$

leading to the eigenvalue and eigenvector equation:

$$(K - \omega^2 M)\, x = 0 \tag{8.4}$$

the solutions of which are the N eigenmodes and the N associated circular eigenfrequencies ω_i of the system, N being the number of degrees of freedom.

Mechanisms and rigid body modes are the m non trivial solutions of the equation

$$K u = 0 \tag{8.5}$$

and may be considered as particular eigenmodes with corresponding zero eigenvalues.

The following important orthogonality relations hold between the different normal modes:

$$u'_{(r)} M u_{(j)} = \mu_j \delta_{rj} \qquad (j = 1, \ldots, m)$$

$$K u_{(j)} = 0 \qquad (r = 1, \ldots, m)$$

$$x'_{(s)} M u_{(j)} = 0 \qquad (j = 1, \ldots, m) \quad (s = m+1, \ldots, N) \tag{8.6}$$

$$x'_{(s)} M x_{(i)} = \mu_i \delta_{si} \qquad (i = m+1, \ldots, N)$$

$$x'_{(s)} K x_{(i)} = \gamma_i \delta_{si} \qquad (s = m+1, \ldots, N)$$

where μ_i and γ_i are the generalized mass and stiffness, respectively, associated with the ith normal mode shape and are related by the relation:

(8.7)
$$\gamma_i = \omega_i^2 \mu_i$$

δ_{ij} being the Kronecker's symbol.

If the system has no mechanisms nor rigid body degrees of freedom, its stiffness matrix K is non singular and equation (8.4) may be rewritten as:

(8.8)
$$K^{-1} M x_{(i)} = \frac{1}{\omega_i^2} x_{(i)}$$

showing that the eigenmodes $x_{(i)}$ are the eigenvectors of the matrix

(8.9)
$$D = K^{-1} M$$

associated with the eigenvalues

$$\nu_i = \frac{1}{\omega_i^2}$$

The matrix D , which plays a very important role in structural dynamics, is called the dynamic flexibility matrix of the system. The extension of the motion of dynamic flexibility matrix to hypostatic systems was first carried out by Fraeijs de Veubeke [1] who demonstrated that the extended dynamic flexibility matrix of an hypostatic system is given by:

$$D = G M \tag{8.10}$$

with

$$G = A G_{iso} A' \tag{8.11}$$

where

$$A = \left(E - \sum_{j=1}^{m} \frac{u_{(j)} u'_{(j)} M}{u'_{(j)} M u_{(j)}} \right) \tag{8.12}$$

is a projection matrix with selective properties:

$$\begin{aligned} A u_{(r)} &= 0 \qquad (r = 1, \dots, m) \\ A x_{(s)} &= x_{(s)} \qquad (s = m+1, \dots, N). \end{aligned} \tag{8.13}$$

G_{iso} denotes a symmetrical static flexibility matrix obtained by imposing to the system any isostatic reference frame. Assuming the symmetrical form, or equivalently, the orthogonality to all zero frequency modes, yields the particular extended static flexibility matrix G given by (8.11).

The following pseudo-inversion relations hold between the K and G matrices:

$$\begin{aligned} K G &= A' \\ G K &= A \end{aligned} \tag{8.14}$$

Finally, equation (8.8) becomes, for any hypostatic system:

$$K\, u_{(j)} = 0 \qquad (j = 1, \ldots, m)$$

(8.15)

$$G\, M x_{(i)} = \frac{1}{\omega_i^2}\, x_{(i)} \qquad (i = m+1, \ldots, N)$$

The mode displacement method

The orthogonality relations (8.6) show that the eigenmodes $(u_{(j)}, x_{(i)})$ are linearly independent and form a complete set in which the general solution of basic equation (8.1) admits a unique expansion:

$$q = \sum_{j=1}^{m} \zeta_j(t)\, u_{(j)} + \sum_{i=m+1}^{N} \eta_i(t)\, x_{(i)} \tag{8.16}$$

so that (8.1) can be rewritten as:

$$\sum_{j=1}^{m} \ddot{\zeta}_j M u_{(j)} + \sum_{i=m+1}^{N} (\ddot{\eta}_i M x_{(i)} + \eta_i K x_{(i)}) = f(t) \tag{8.17}$$

Successive premultiplication of (8.17) by $(u'_{(r)}, r = 1, \ldots, m)$, $(x_{(s)}, s = m+1, \ldots, N)$, with use of the orthogonality relations (8.6), yields then the uncoupled normal euqations of motion:

$$\ddot{\xi}_r = \frac{u'_{(r)}\, f(t)}{\mu_r} \qquad (r = 1, \ldots, m)$$
$$\ddot{\eta}_s + \omega_s^2 \eta_s = \frac{x'_{(s)}\, f(t)}{\mu_s} \qquad (s = m+1, \ldots, N) \tag{8.18}$$

which may be easily solved, by use of the Laplace transform and convolution theorem, into:

$$\xi_r = \frac{u_{(r)}}{\mu_r} \int_0^t (t-\tau)\, f(\tau)\, d\tau \qquad (s = 1, \ldots, m)$$
$$\eta_s = \frac{x_{(s)}}{\omega_s \mu_s} \int_0^t \sin \omega_s (t-\tau)\, f(\tau)\, d\tau \qquad (s = m+1, \ldots, N) \tag{8.19}$$

providing zero initial conditions are assumed:

$$\xi_r(0) = \dot{\xi}_r(0) = 0 \qquad (r = 1, \ldots, m)$$
$$\eta_s(0) = \dot{\eta}_s(0) = 0 \qquad (s = m+1, \ldots, N). \tag{8.20}$$

Hence, the general solution of (8.1) is given by:

$$q(t) = \sum_{j=1}^{m} \frac{u_{(j)}\, u'_{(j)}}{\mu_j} \int_0^t (t-\tau)\, f(\tau)\, d\tau$$
$$+ \sum_{i=m+1}^{N} \frac{x_{(i)}\, x'_{(i)}}{\omega_i\, \mu_i} \int_0^t \sin \omega (t-\tau)\, f(\tau)\, d\tau \tag{8.21}$$

And the equivalent static force which would at each time, produce the actual stress field is:

$$g(t) = K\, q_{(t)} = \sum_{i=m+1}^{N} \frac{M\, x_{(i)}\, x'_{(i)}}{\mu_i}\, \omega_i \int_0^t \sin \omega_i (t-\tau)\, f(\tau)\, d\tau$$

(8.22)

Equations (8.21) and (8.22) give the complete solution of basic equation (8.1) providing all the normal eigenmode-shapes of the system are known. If this is not the case, and it will never occur practically with large systems for which a complete modal analysis would be prohibitive, only the first n normal modes are known and can be used to compute dynamic response.

The problem is then to determine under which conditions an incomplete n-fold set of low frequency normal modes will be sufficient for computing of displacements and stresses with satisfactory accuracy by truncation of the expansions (8.21) and (8.22) to their n first terms. To this purpose, let us suppose the transient external loads are of the type:

(8.23) $$f(t) = p \cdot \Phi(t)$$

where p is a constant vector modulated by time history $\Phi(t)$ (it should be noted that this is not a restriction since any general load can be expressed by a superposition of such terms). He have now to check the convergence of the expressions:

$$\sum_{i=m+1}^{n} \frac{x_{(i)}\, x'_{(i)}\, p}{\mu_i} \frac{1}{\omega_i} \int_0^t \sin \omega_i (t - \tau)\, \Phi(\tau) d\tau \qquad (8.24)$$

and

$$\sum_{i=m+1}^{n} \frac{M\, x_{(i)}\, x'_{(i)}\, p}{\mu_i}\, \omega_i \int_0^t \sin \omega_i (t - \tau)\, \Phi(\tau) d\tau \qquad (8.25)$$

when n is increased. It can be seen that each term of the expressions (8.24) and (8.25) is a product of two different terms: a static term:

$$\frac{x_{(i)}\, x'_{(i)}\, p}{\mu_i} \qquad \text{or} \qquad \frac{M\, x_{(i)}\, x'_{(i)}\, p}{\mu_i}$$

which depends only on the static force p and a time dependent spectral term:

$\frac{1}{\omega_i} \sin \omega_i * \Phi(t)$ or $\omega_i \sin \omega_i t * \Phi(t)$ which depends only which depend only of the spectrum of the time history $\Phi(t)$. Hence two types of convergence can be expected:

a) a pseudo-static convergence which will oscur if n is large enough to allow a good representation of the static force p by means of this expansion in the n fold set of the known modes, that is to say that p must be orthogonal to the $(N - n)$ unknown modes which will consequently not be existe.

b) A spectral convergence which will occur if the time history $\Phi(t)$ is such that the convolution integrals

$\frac{1}{\omega_i} \sin \omega_i t * \Phi(t)$ or $\omega_i \sin \omega t * \Phi(t)$ converge to zero when i is increased so that the component of the response in the unknown high frequency modes may be neglected in comparison with that in the known low-frequency modes

If we consider for instance the step loading

$$\Phi(t) = \nu(t)$$

$$\frac{\sin \omega_i t * \Phi(t)}{\omega_i} = \frac{1 - \cos \omega_i t}{\omega_i^2}$$

we obtain a spectral convergence with rate $\frac{1}{\omega_i^2}$ for the displacements but no spectral convergence for the equivalent static force and hence no spectral convergence for the stresses.

Let us now consider an harmonic loading:

$$\Phi(t) = \sin \omega t$$

$$\frac{\sin \omega_i t * \Phi(t)}{\omega_i} = \frac{\omega_i \sin \omega t - \omega \sin \omega_i t}{\omega_i (\omega_i^2 - \omega^2)}$$

provided ω_i is great compared to ω we get again a spectral convergence with rate $\frac{1}{\omega_i^2}$ for the displacements and, again no spectral convergence for the stresses.

It should be noted that, while the quasi-static convergence is valid for both displacements and stresses calculation, spectral convergence of the displacement solution does not at all imply convergence for the stresses.

If convergence is insured, we may write the mode displacement approximations for displacements and equivalent static force, which are:

$$q(t) = \sum_{j=1}^{m} \frac{u_{(j)} u'_{(j)}}{\mu_j} \int_0^t (t-\tau)\, f(\tau)\, d\tau$$

$$+ \sum_{i=m+1}^{n} \frac{x_{(i)} x'_{(i)}}{\omega_i \mu_i} \int_0^t \sin \omega_i (t-\tau)\, f(\tau) d\tau \qquad (8.26)$$

$$g(t) = \sum_{i=m+1}^{n} \frac{x_{(i)} x'_{(i)}}{\mu_i}\, \omega_i \int_0^t \sin \omega_i (t-\tau)\, f(\tau) d\tau \qquad (8.27)$$

The mode acceleration method

This method, generally attributed to Williams [2] (see also [3] , [4]), was in fact, already presented by Lord Rayleigh [5] (V, 100) in a slightly different, but equivalent manner.

Basic equation (8.1) is a dynamic equilibrium equation in which three different forces are involved:

the applied force : $f(t)$

the equivalent static force : $g = Kq$

the inertia forces : $a = M\ddot{q}$

If inertia forces were removed from the system, the solution of (8.1) would be the quasi-static solution, which is, with assumed nonsingular stiffness matrix:

(8.28) $$q(t) = K^{-1} F(t)$$

and can thus be obtained independently of any modal analysis. If inertia forces are to be taken into consideration the exact solution is obtained by adding to (8.28) the static displacements they produce:

(8.29) $$q(t) = K^{-1} F(t) - K^{-1} a$$

Thus, as pointed out by Williams, only the inertia forces need to be expanded in the incomplete n fold set of the known modes, but neither the applied force nor the equivalent static force, as done is the mode displacement method. We shall now write according to (8.16)

(8.30) $$q(t) = K^{-1} \left\{ F(t) - \sum_{i=1}^{n} \ddot{\eta}_i \, M \, x_{(i)} \right\}$$

Using (8.18) and (8.19) we get:

(8.31) $$\ddot{\eta}_i = \frac{x'_{(i)}}{\mu_i} \left\{ F(t) - \omega_i \int_0^t \sin \omega_i (t-\tau) \, F(\tau) \, d\tau \right\}$$

and the new approximate solution of (8.1) becomes:

$$q(t) = K^{-1} \left\{ F(t) - \sum_{i=1}^{n} \frac{M \, x_{(i)} x'_{(i)}}{\mu_i} \left[F(t) - \omega_i \int_0^t \sin(t-\tau) F(\tau) d\tau \right] \right\}$$
(8.32)

that is with use of (8.8)

$$q(t) = \sum_{i=1}^{n} \frac{x_{(i)} x'_{(i)}}{\omega_i \mu_i} \int_0^t \sin \omega_i (t-\tau) F(\tau) d\tau + \left\{ K^{-1} - \sum_{i=1}^{n} \frac{x_{(i)} x'_{(i)}}{\omega_i^2 \mu_i} \right\} F(\tau) \tag{8.33}$$

In a quite analogous way, Lord Rayleigh started from the uncoupled normal equations of motion (8.18) and pointed out that the normal inertia force may be neglected in comparison with the normal stiffness force, when the period of the considered mode becomes small when compared with that of the operating external loads. If this is true, except in the n first modes, we may rewrite the normal equations (8.18) in the form:

$$\begin{aligned} \ddot{\eta}_i + \omega_i^2 \eta_i &= \frac{x'_{(i)} F(t)}{\mu_i} & (i = 1, \dots, n) \\ \omega_i^2 \eta_i &= \frac{x'_{(i)} F(t)}{\mu_i} & (i = m+1, \dots, N) \end{aligned} \tag{8.34}$$

thus we get:

$$\begin{aligned} \eta_i &= \frac{x_{(i)}}{\omega_i \mu_i} \int_0^t \sin \omega_i (t-\tau) F(\tau) d\tau & (i = 1, \dots, n) \\ \eta_i &= \frac{x'_{(i)} F(t)}{\omega_i^2 \mu_i} & (i = n+1, \dots, N) \end{aligned} \tag{8.35}$$

and the approximate solution of (8.1) is:

$$q(t) = \sum_{i=1}^{n} \frac{x_{(i)} x'_{(i)}}{\omega_i \mu_i} \int_0^t \sin \omega_i (t-\tau) f(\tau) d\tau + \left\{ \sum_{i=n+1}^{N} \frac{x_{(i)} x'_{(i)}}{\omega_i^2 \mu_i} \right\} f(t)$$

(8.36)

Equivalence of (8.33) and (8.36) is obvious since the stiffness matrix K is equal to its spectral expansion in the complete N fold set of the normal modes:

(8.37) $$K = \sum_{i=1}^{N} \frac{K x_{(i)} x'_{(i)} K}{x'_{(i)} K x_{(i)}}$$

and was assumed to be non singular. Thus, pre-and post- multiplications of (8.37) by K^{-1} , with use of relations (8.6) and (8.7) yields:

(8.38) $$K^{-1} = \sum_{i=1}^{N} \frac{x_{(i)} x'_{(i)}}{\omega_i^2 \mu_i}$$

hence

(8.39) $$K^{-1} - \sum_{i=1}^{n} \frac{x_{(i)} x'_{(i)}}{\omega_i^2 \mu_i} = \sum_{i=n+1}^{N} \frac{x_{(i)} x'_{(i)}}{\omega_i^2 \mu_i}$$

Rayleigh's approach, which does not involve the static flexibility matrix K^{-1} , is directly extensible to hypostatic systems with singular stiffness matrix K , and we get in this case:

$$q(t) = \sum_{j=1}^{m} \frac{u_{(j)} u'_{(j)}}{\mu_j} \int_0^t (t-\tau) f(\tau) d\tau +$$

$$+\sum_{i=m+1}^{n} \frac{x_{(i)}\, x'_{(i)}}{\omega_i\, \mu_i} \int_0^t \sin \omega_i (t-\tau) f(\tau) d\tau +$$

$$+\left\{ \sum_{i=n+1}^{N} \frac{x_{(i)}\, x'_{(i)}}{\omega_i^2\, \mu_i} \right\} f(t) \qquad (8.40)$$

But we may now write, remembering the definition of the extended static flexibility matrix G given in section I and using the analogy between equations (8.8) and (8.15):

$$G = \sum_{i=m}^{N} \frac{x_{(i)}\, x'_{(i)}}{\omega_i^2\, \mu_i} \qquad (8.41)$$

So that the most general expression for the mode acceleration approximate solution of basic equation (8.1) with n known normal modes is:

$$q(t) = \sum_{j=1}^{m} \frac{u_{(j)}\, u'_{(j)}}{\mu_j} \int_0^t (t-\tau) f(\tau)\, d\tau +$$

$$+\sum_{i=m+1} \frac{x_{(i)}\, x'_{(i)}}{\omega_i\, \mu_i} \int_0^t \sin \omega_i (t-\tau) f(\tau) d\tau +$$

$$+\left\{ G - \sum_{i=m+1}^{n} \frac{x_{(i)}\, x'_{(i)}}{\omega_i^2\, \mu_i} \right\} f(t) \qquad (8.42)$$

The equivalent pseudo-static force being now given by:

$$g(t) = \sum_{i=m+1}^{n} \frac{M\, x_{(i)}\, x'_{(i)}}{\mu_i}\, \omega_i \int_0^t \sin \omega_i (t-\tau) f(\tau) d\tau +$$

$$+\left(A' - \sum_{i=m+1}^{n} \frac{M\, x_{(i)}\, x'_{(i)}}{\mu_i} \right) f(t) \qquad (8.43)$$

(pseudo-static because $A'f(t)$ contains the additional inertia forces involved by the rigid body accelerations).

If we compare the results of the mode displacement method (8.26) with those of the mode acceleration method (8.42), we see that they differ by the term:

$$s(t) = \left\{ G - \sum_{i=m+1}^{n} \frac{x_{(i)}\, x'_{(i)}}{\omega_i^2\, \mu_i} \right\} f(t) \tag{8.44}$$

which is the product of the applied load by the deflected extended flexibility matrix and represents the pseudo-static displacements in the $(N-n)$fold set of the unknown modes where inertia forces have been neglected.

The mode acceleration method will generally give better convergence than the mode displacement method since the expansion of the single inertia force can be expected to converge more rapidly than that of the whole operating forces. It can be seen that there is no longer problem with pseudo-static convergence since the method is statically exact. Furthermore, the mode acceleration correction (8.44) provides a useful criterion to check the pseudo-static convergence of the mode displacement method. In order to check the spectral convergence of the method, let us again consider the step loading:

$$\Phi(t) = \nu(t)$$

$$\frac{\sin \omega_i t * \Phi(t)}{\omega_i} = \frac{1 - \cos \omega_i t}{\omega_i^2}$$

it can be seen that in an unknown mode (i) the exact contribution $\frac{1-\cos\omega_i t}{\omega_i^2}$ which has been neglected in the mode displacement method is now replaced by $\frac{1}{\omega_i^2}$ which represents its mean value over a period, but no significant improvement of the spectral convergence has been obtained.

For harmonic loading:

$$\Phi(t) = \sin \omega t$$

$$\frac{\sin \omega_i t * \Phi(t)}{\omega_i} = \frac{\omega_i \sin \omega t - \omega \sin \omega_i t}{\omega_i(\omega_i^2 - \omega^2)}$$

the difference between exact contribution of the unknown mode (i) and the assumed contribution $\frac{\sin \omega t}{\omega_i^2}$ is:

$$\frac{\omega}{\omega_i^2} \; \frac{\omega \sin \omega t - \omega_i \sin \omega_i t}{\omega_i^2 - \omega^2}$$

so that the spectral convergence rate is now $\frac{1}{\omega_i^3}$ for the displacements, hence $\frac{1}{\omega_i}$ for the stresses, provided ω_i is high enough compared with ω .

Thus, in the harmonic case, we get a significant improvement of the spectral convergence when using the mode acceleration method.

Numerical results

Several examples have been chosen in order to illustrate the methods described here, including an extension of the mode acceleration method to systems with small orthogonal viscous damping (see appendix I). Results are compared with those of Houbolt's direct recurrence method [6] (see appendix II) or those of an exact analytical method. The modally uncoupled precise integration operator recently presented by Dunham, Nickell and Stickler [7] (see appendix III) has been checked and was found to be of great interest when analytical integration of the normal equations of motion (8.18) becomes prohibitive.

The first example deals with a three dimensional frame submitted to a concentrate step load which is removed after a short time $T = 10^{-2}$ sec. The frame (see figure 1) has been idealized with sixteen Euler's beam element of the third degree leading to a discrete system with fourty-eight degrees of freedom. Since the load is applied during a very short time, only the mode displacement method is suitable but the pseudo-static convergence can be checked by means of the mode acceleration correction (8.44). Figures 1 and 2 show the time histories of the most representative displacement and bending moment obtained by the mode displacement method, with twelve modes, and by the Houbolt's method with time step $\Delta t = 5.10^{-4}$ sec. Agreement between the two methods is excellent. Figure 3 illustrates the convergence of the maximum strain energy (occurring at time $t = 0.035$ sec.) versus time step Δt for the Houbolt's method and versus the number n of used modes for the mode displacement method: stabilization of the mode displacement solu-

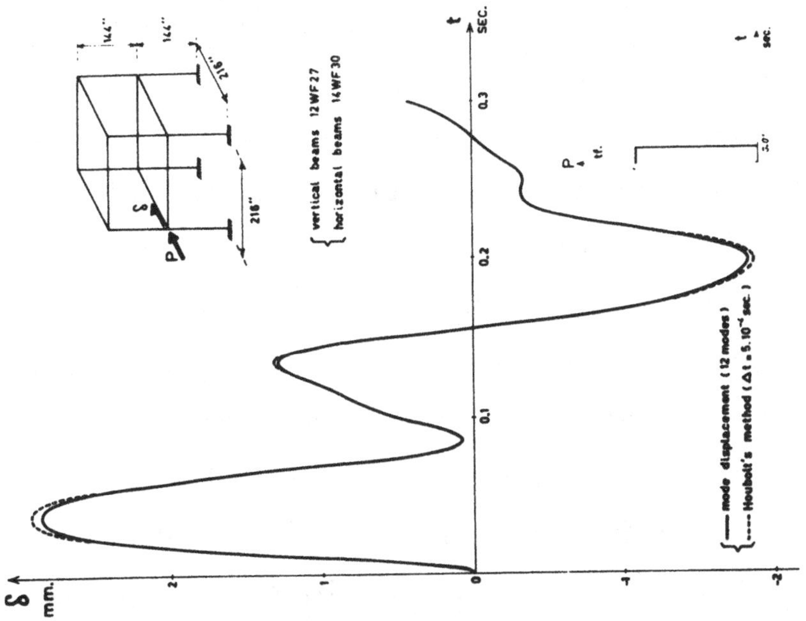

Fig. 1. Three dimensional frame under impulse load time history of the loading point deflection.

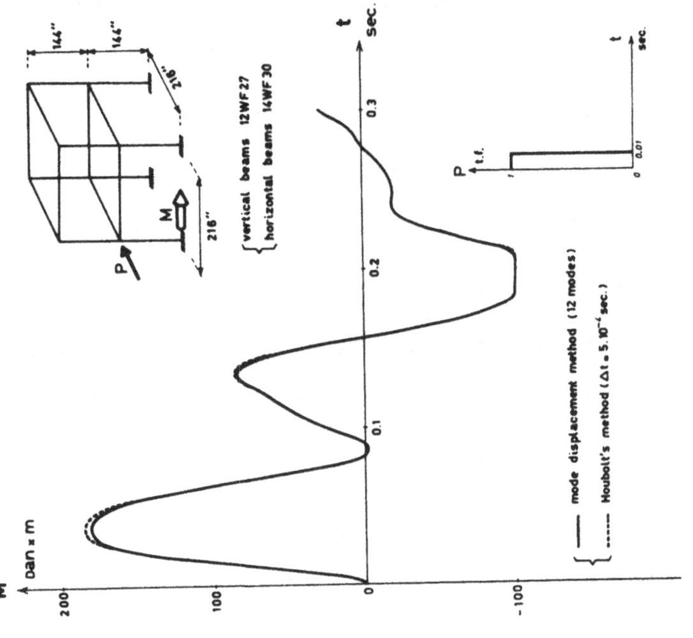

Fig. 2. Three dimensional frame under impulse load time history of the moment M.

tion with twelve modes and progressive bracketting when n is increased and Δt decreased are to be outlined.

The second example consists in a uniform pinned-beam overcrossed by a concentrated load P with constant velocity v. Exact analytical solution has been computed by expansion of Navier's eigenfunctions and finite element analysis has been carried out by idealizing the beam with twenty Euler's beam of the third degree leading to a system with fourty degrees of freedom. Discretization of the travelling load and integration of the normal equations of motion are somewhat laborious to be detailed here. A velocity equal to the second critical speed has been assumed. Figures 4 and 5 show the time history of the mid span deflection and bending moment while figures 6 and 7 give the deflection and bending moment distribution at the maximum mid span deflection time. All these results were computed analytically, by the Houbolt's method with time step $\Delta t = 5\,10^{-4}$ sec., by the mode displacement method and by the mode acceleration method with the five first normal modes. Excellent agreement between the different methods can be again noticed. Furthermore, this problem has been chosen to check the modally uncoupled precise integration operator of Dunham, Nickell and Stickler: integration of the normal equations of motion by means of this operator was carried out with time step $\Delta t = 5\,10^{-4}$ sec. and no significant difference with analytical integration could be noticed over a zero to 0.5 sec. time

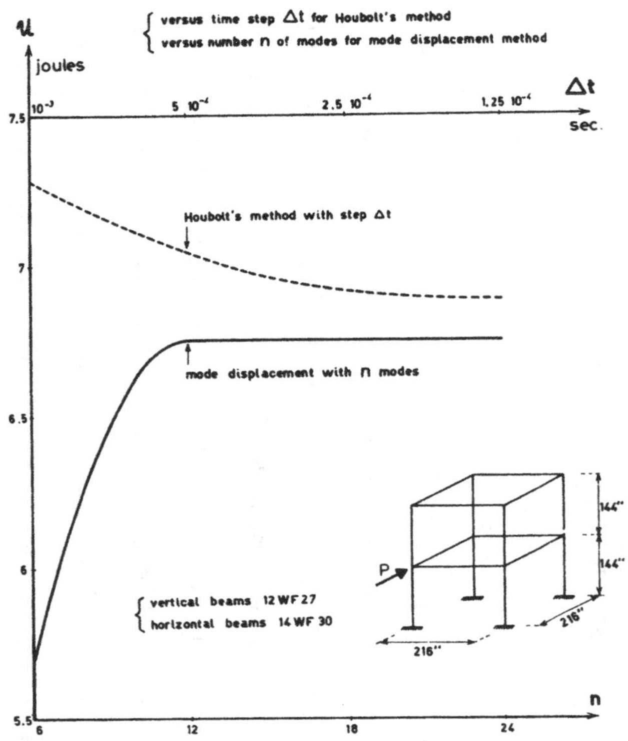

Fig. 3. Three dimensional frame under impulse load convergence of the maximum strain energy at time t = 0.035 sec.

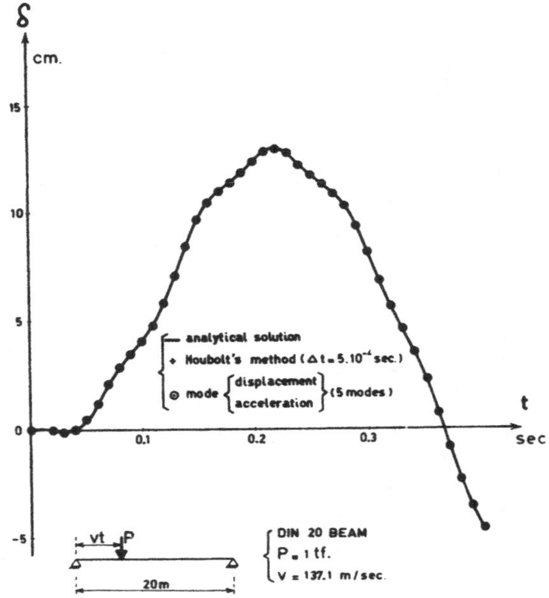

Fig. 4. Pinned-pinned beam under a load travelling at 2nd critical speed time-history of the mid-span deflection.

range.

The third example illustrates the extension of the mode acceleration method to hypostatic systems by means of a free-free beam under mid-span concentrated step loading. Analytical solution was obtained by expansion in Duncan's eigenfunctions while finite element analysis was made with twenty Euler's beam element of the third degree giving a system with forty two degrees of freedom. Mode displacement and mode acceleration solutions were computed with the five first symmetric modes. Figure 8 shows the mid-span deflection history over three periodes and Figure 9 the mid-span bending moment history over the first period. Excellent agreement for the deflection and satisfactory bracketting of the bending moment can be noticed.

The fourth example is a simply supported square plate under mid point harmonic excitation with 3% assumed orthogonal viscous damping ratio. Analytical solution was obtained by expansion in Navier's eigenfunctions. Finite elements idealization was made with sixteen Kirchhoff's plate elements of the third degree for a quater of the plate leading to a system with eighty degrees of freedom. Amplitude and phase of the forced harmonic response were computed by the mode acceleration method, with eleven symmetric-symmetric modes, for several excitation frequencies including resonance frequencies. Figure 10 shows the frequency history of the mid

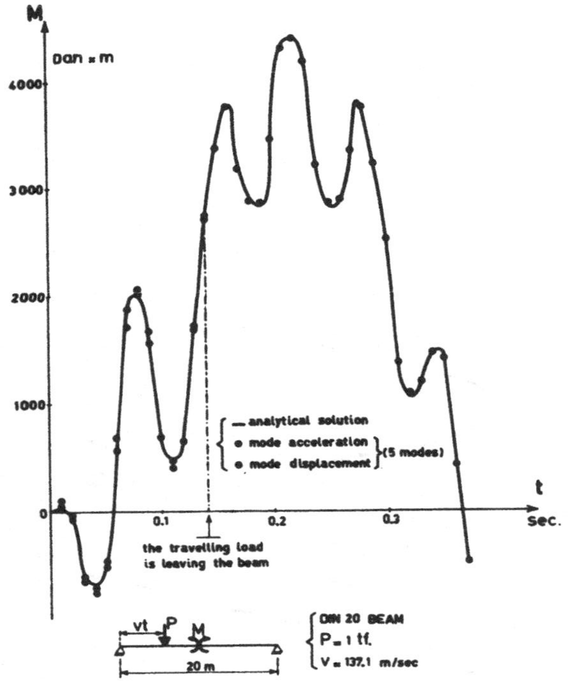

Fig. 5. Pinned-pinned beam under a load travelling at 2nd critical speed time-history of the mid-span nending moment.

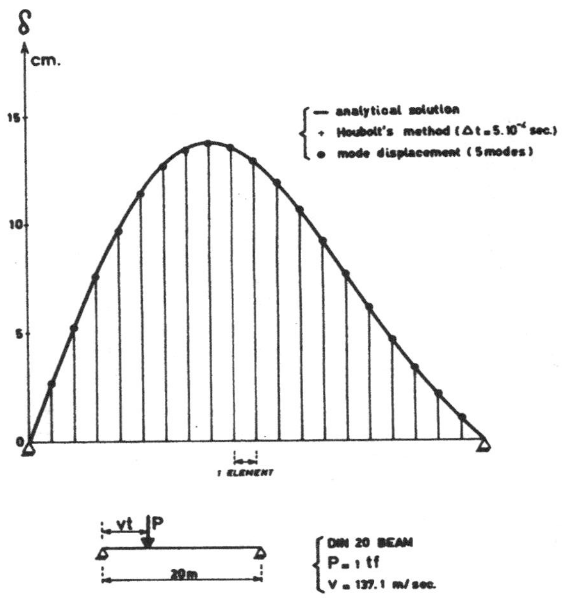

Fig. 6. Pinned-pinned beam under a load travelling at 2nd critical speed beam deflection at time t = 0.22 sec..

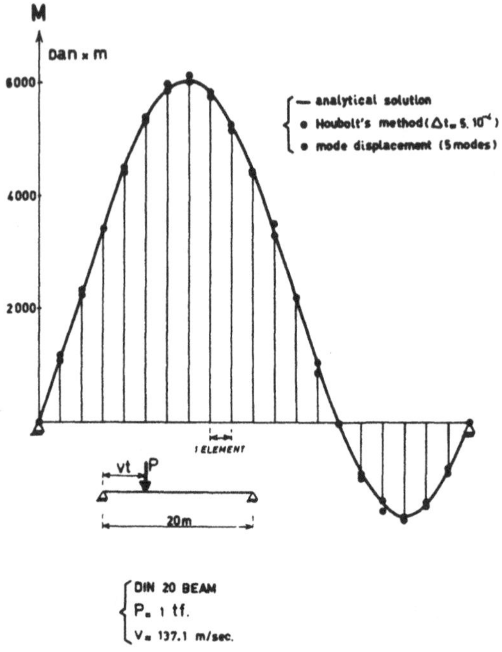

Fig. 7. Pinned-pinned beam under a load travelling at 2nd critical speed bending moment distribution at time t = 0.22 sec.

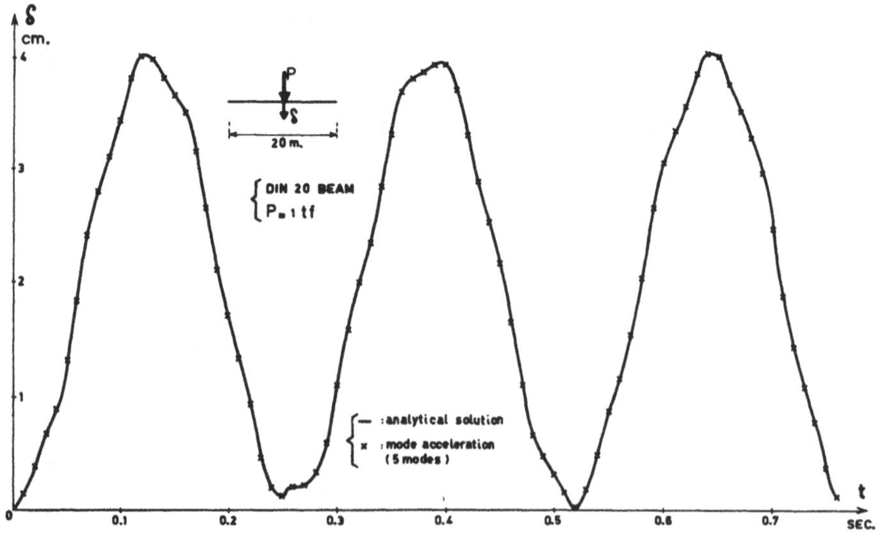

Fig. 8. Free-free beam under mid span step loading time history of the mid-span deflection

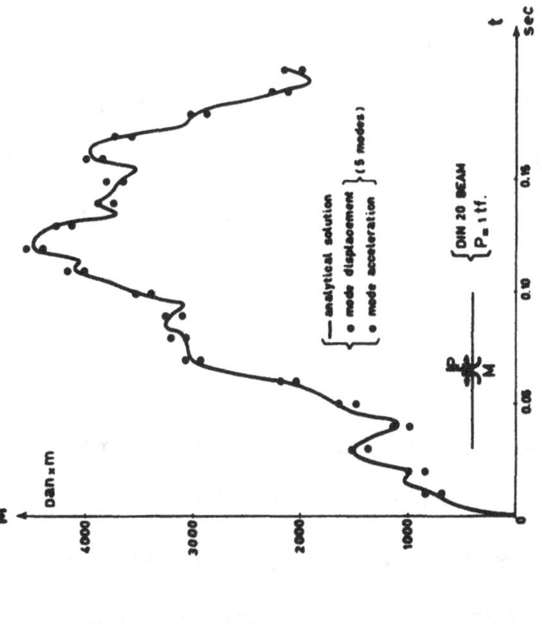

Fig. 9. Free-free beam under mid-span step-loading time history of the mid-span bending moment.

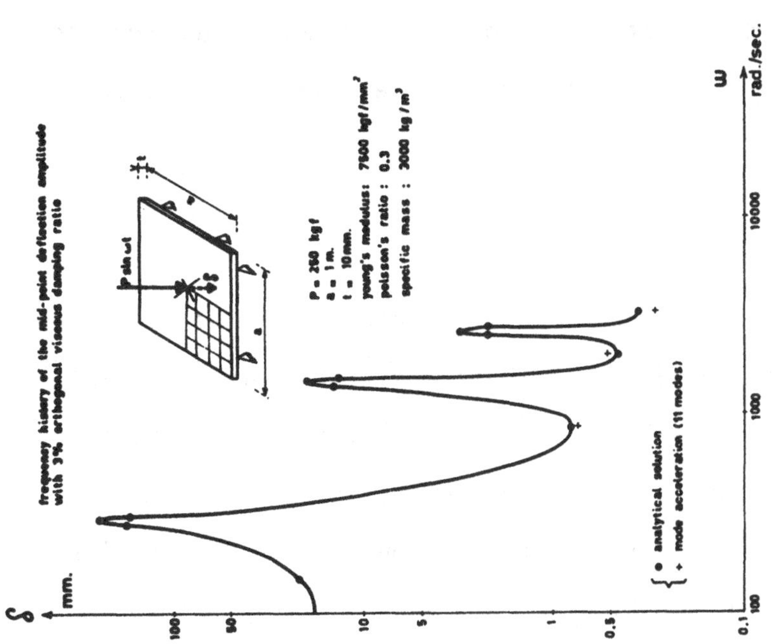

Fig. 10. Simply supported square plate under concentrated harmonic excitation at the mid-point.

point deflection amplitude up to the third resonance peak and outlines again the good accuracy of the mode acceleration method.

Conclusion

As announced in the beginning methods based on modal analysis are seen to give very satisfactory transient response although only a generally small number of significant low frequency modes needs to be taken into consideration. Both mode displacement and mode acceleration methods have a simple finite element formulation and are easily extensible to singular systems.

The mode acceleration method was seen to be a simple improvement of the mode displacement method by means of the quasi-static contribution of the unknown modes which provides also a suitable criterion for checking the convergence: is this contribution small, so both methods are valid with advantage when using the mode acceleration method. Is this contribution not small so the number of known modes is unsufficient and both methods fail since the fact that the mode acceleration method will generally give a better result that the mode displacement method does no longer imply that this result will be good. When comparing the mode displacement or the mode acceleration method with direct methods, like Houbolt's method, it is seen that a substantial economy is due to the fact that only n second order linear differential equations are to be solved instead of iterating the complete $(N \times N)$ coupled system. Of course, the cost of the modal analysis must be taken into consideration but it must be kept in mind that a single modal analysis allows the treatment of any number of response problems.

REFERENCES

[1] B.M. Fraeijs de Veubeke, "Iteration in semidefinite eigenvalue problems", J.A.S., vol. 22, N° 10, October 1955

[2] D. Williams, "Displacements of a Linear Elastic System under a Given Transient load", The Aeronautical Quaterly, vol. 1, August 1949

[3] R.L. Bisplinghoff, G. Isakson, T.H.H. Pian, "Methods in Transient Stress Analysis", J.A.S., vol. 17, n° 5, May 1950.

[4] J.W. Mar, T.H.H. Pian, J.M. Calligeros, "A note on methods for the determination of transient stresses", J.A.S., January 1956.

[5] J.W.S. Rayleigh, "The theory of Sound",(Volume I) Dover, New York, 1945.

[6] J.C. Houbolt, "A recurrence matrix solution for the dynamic response of elastic aircraft", J.A.S., vol. 17, n° 9, September 1950.

[7] R.S. Dunham, R.E. Nickell, D.C. Stickler, "Integration operators for transient structural response, Computers and Structures", vol. 2, n° 1/2, February 1972.

Appendix I

If viscous damping is assumed, basic equation (8.1) becomes:

$$M\ddot{q} + B\dot{q} + Kq = f(t) \tag{8.45}$$

where B is a symmetric non negative damping matrix. Assuming (8.16) leads then to the following coupled normal equations of motion:

$$\mu_r\ddot{\eta}_r + \sum_{i=1}^{N}\left[x'_{(r)}\,B\,x_{(i)}\right]\dot{\eta}_i + \gamma_r\,\eta_r = x'_{(r)}f(t) \tag{8.46}$$

These equations show that damping generally couples the normal modes of the associated undamped system. If damping is small and if coupling of modes may be neglected a great simplification of (8.46) occurs when assuming the following additional orthogonality relations:

$$x'_{(r)}\,B\,x_{(i)} = 2\,\varepsilon_r\,\omega_r\,\mu_r\,\delta_{ri} \tag{8.47}$$

where ε_i is the ratio of critical damping in the ith mode. It must be noted that hypothesis (8.47) is rather restrictive, since neglecting coupling of the modes, and is to be considered only as an occasionally valid approximation.

If (8.47) be assumed, (8.46) becomes:

$$\ddot{\eta}_r + 2\varepsilon_r \omega_r \dot{\eta} + \omega_r^2 \eta_r = \frac{x'_{(i)} F(t)}{\mu_r} \tag{8.48}$$

the solution of which is, with assumed zero initial conditions:

$$\eta_r(t) = \frac{x'_{(r)}}{\mu_r \sqrt{1-\varepsilon_r^2}\,\omega_r} \int_0^t e^{-\varepsilon_r \omega_r (t-\tau)} \sin\left[\sqrt{1-\varepsilon_r^2}\,\omega_r (t-\tau)\right] F(\tau)\,d\tau \tag{8.49}$$

leading to the following mode displacement response:

$$q(t) = \sum_{i=1}^{n} \frac{x_{(i)} x'_{(i)}}{\mu_i \sqrt{1-\varepsilon_i^2}\,\omega_i} \int_0^t e^{-\varepsilon_i \omega_i (t-\tau)} \sin\left[\sqrt{1-\varepsilon_i^2}\,\omega_i (t-\tau)\right] F(\tau)\,d\tau \tag{8.50}$$

Now, if damping is small, both inertia and damping forces may be neglected for higher modes when compared to the corresponding stiffness and applied forces. Thus, by analogy with the mode acceleration method for undamped systems, we may use here a mode velocity method and improve (8.50), by means of the quasi-static contribution of the unknown modes, into:

$$q(t) = \sum_{i=1}^{n} \frac{x_{(i)} x'_{(i)}}{\mu_i \sqrt{1-\varepsilon_i^2}\,\omega_i} \int_0^t e^{-\varepsilon_i \omega_i (t-\tau)} \sin\left[\sqrt{1-\varepsilon_i^2}\,\omega_i (t-\tau)\right] F(\tau)\,d\tau + \left[K^{-1} \sum_{i=1}^{n} \frac{x_{(i)} x'_{(i)}}{\omega_i^2 \mu_i}\right] F(t) \tag{8.51}$$

Appendix II

If a cubic curve is assigned to pass through four successive ordinates with abscissa step h , the following difference equation holds:

$$\ddot{q}_n = \frac{1}{h^2}(2q_n - 5q_{n-1} + 4q_{n-2} - q_{n-3}) \tag{8.52}$$

Setting (8.52) into basic equation (8.1) with time step $\Delta t = h$ yields the Houbolt's backwards difference algorithm (see reference)

$$\left(M + \frac{h^2}{2}K\right)q_n = \frac{h^2}{2}f_n + M\left(\frac{5}{2}q_{n-1} - 2q_{n-2} + \frac{1}{2}q_{n-3}\right) \tag{8.53}$$

which is unconditionally stable but subject to spurious damping.

Iteration must be started with a special procedure using the derivatives at the third of the four points:

$$\begin{aligned} \dot{q}_n &= \frac{1}{6h}(2q_{n+1} + 3q_n - 6q_{n-1} + q_{n-2}) \\ \ddot{q}_n &= \frac{1}{h^2}(q_{n+1} - 2q_n + q_{n-1}) \end{aligned} \tag{8.54}$$

Hence, with zero initial conditions:

$$\begin{aligned} q_0 &= 0 \\ \dot{q}_0 &= 0 \end{aligned} \tag{8.55}$$

we get the following starting procedure

(8.56)
$$q_1 = \frac{h^2}{6}\left(M + \frac{h^2}{6}K\right)^{-1}(f_1 + 2f_0)$$

$$q_0 = 0$$

$$q_{-1} = h^2 M^{-1} f_0 - q_1$$

Appendix III

When numerical integration of the normal equations of motion (8.18) is required, the following modally uncoupled precise integration operator, due to Dunham, Nickell and Stickler (see reference) can be successfully used:

$$\begin{bmatrix} \eta_i\,(t+\Delta t) \\ \dot{\eta}_i\;\; t+\Delta t) \\ \ddot{\eta}_i\;\; t+\Delta t) \end{bmatrix} = \Lambda(\omega_i\,,\,\Delta t)\cdot\begin{bmatrix} \eta_i\,(t) \\ \dot{\eta}_i(t) \\ \ddot{\eta}_i\,(t) \end{bmatrix} + g(\omega_i\,,\,\Delta t)\cdot\frac{x'_{(i)}\,f(t+\Delta t)}{\mu_i}$$

with

$$A(\omega_i\,,\Delta t) = \begin{bmatrix} \dfrac{\sin(\omega_i\Delta t)}{\omega_i\Delta t} & \dfrac{\sin(\omega_i\Delta t)}{\omega_i} & \dfrac{1}{\omega_i^2}\left[\dfrac{\sin(\omega_i\Delta t)}{\omega_i\Delta t} - \cos(\omega_i\Delta t\right] \\[2ex] \dfrac{\cos(\omega_i\Delta t)-1}{\Delta t} & \cos(\omega_i\Delta t) & \dfrac{1}{\omega_i}\left[\sin(\omega_i\Delta t) + \dfrac{\cos(\omega_i\Delta t)-1}{\omega_i\Delta t}\right] \\[2ex] -\;\dfrac{\omega_i\,\sin(\omega_i\Delta t)}{\Delta t} & -\,\omega_i\sin(\omega_i\Delta t) & \cos(\omega_i\Delta t) - \dfrac{\sin(\omega_i\Delta t)}{\omega_i\Delta t} \end{bmatrix}$$

and

$$g(\omega_i, \Delta t) = \begin{bmatrix} \dfrac{1}{\omega_i^2}\left[1 - \dfrac{\sin(\omega_i \Delta t)}{\omega_i \Delta t}\right] \\ \dfrac{1}{\omega_i^2 \Delta t}\left[1 - \cos(\omega_i \Delta t)\right] \\ \dfrac{\sin(\omega_i \Delta t)}{\omega_i \Delta t} \end{bmatrix}$$

This operator is unconditionally stable with no spurious damping or error in vibrational period.

HEAT CONDUCTION

M. HOGGE

B. FRAEIJS de VEUBEKE

Laboratoire de Techniques Aéronautiques
et Spatiales
Université de Liège

P R E F A C E

The steady state heat conduction problem is formulated in terms of the basic physical quantity attached to this kind of boundary value problem: a dissipation functional.

Dual minimum principles are derived from which respectively upper and lower bounds to the dissipation functional can be obtained from corresponding models of finite elements.

The temperature and heat flow models are derived and tested from the viewpoint of numerical effectiveness. For the latter, either the heat flow itself is discretized, or this is done through a heat stream function.

Numerical bounds are presented for the dissipation functional of certain problems where both types of models were applied.

Udine, July 1972

PART I

LOWER BOUNDS TO A DISSIPATION FUNCTIONAL BY TEMPERATURE COMPATIBLE FINITE ELEMENTS

I. INTRODUCTION

Finite element methods are best known from their applications to linear elasticity theory. Dual single-field variational principles, like the principle of minimum total energy and the principle of minimum complementary energy, were shown to be advantageous in the construction of mathematical models of finite elements and in the numerical estimation of the accuracy of the approximations [1] , [2] , [3] .

Non structural continuum problems have received less attention and only recently [4] , [5] has the dual formulation been presented for other field problems where variational principles are available.

In the present paper attention is focused on the steady state heat conduction problem, although other boundary value problems, like steady seepage through porous media, electrostatic and electromagnetic fields, isoenergetic flows of ideal fluids can benefit from similar treatment.

The basic dual theorems can be derived either from variational manipulations [5] or from inequalities based on thermal behaviour of the continuum. The latter approach starts from an integrated thermal balance of the continuum and is analogous to the virtual work principle of elasticity. It stresses the fact that the variational character of the theorems is not essential for their use in conjunction with the finite element method. The basic physical quantity of steady state heat conduction, i.e. the dissipation functional, is introduced and the two minimum theorems are derived from its positive definite character. Special emphasis is given on the nature of the transition conditions required when the domain is subdivided into finite subdomains. The dual character of the theorems extends to the benefit of the dissipation functional a numerical estimate of the convergence by upper bounds (temperature approximations) and lower bounds (heat flow approximations) when no heat sinks are prescribed. When the body is in contact with a uniform temperature bath, taken as zero level, the role of bounds is reversed. The general case of a mixed boundary value problem where both sinks and non-uniform external temperature prevail can be treated by superposition if convergence estimates are required.

The following sections are devoted to the construction of mathematical models of finite elements of the compatible temperature type based on the first minimum theorem [6] . The particular concepts of shaping and bubble functions are introduced for the temperature field.

An original treatment of the convection boundary of the continuum is presented which is necessary to obtain correct bounds in a dual analysis. Some numerical examples show practical use of 2D - and axisym - temperature finite elements in solving heat conduction problems.

2. FIELD EQUATIONS AND BOUNDARY CONDITIONS

The heat conducting body occupies a volume D bounded by a surface ∂D , the outward normal of which has direction cosines $n_i (i = 1,2,3)$. Let $T(x_i)$ be the steady but non uniform temperature distribution referred to cartesian coordinates.

The temperature gradient is denoted by

(2.1) $$\partial_i T = \frac{\partial T}{\partial x_i}$$

and heat flow q_i is governed by Fourier's law

(2.2) $$q_j = - k_{ji} \partial_i T$$

where k_{ij} is a symmetrical heat conductivity tensor. When the principal axes of this tensor have constant orientation from point to point and the cartesian axes are taken parallel to them, the simpler law prevails

(2.3) $$q_1 = - k_1 \partial_1 T \qquad q_2 = - k_2 \partial_2 T \qquad q_3 = - k_3 \partial_3 T$$

If Q denotes a heat sink distribution per unit volume, the heat flow balance requires

(2.4) $$\partial_j q_j + Q = 0$$

Boundary conditions of various type are considered. If the body is in contact with a heat bath, whose temperature can be maintened at a prescribed level T_e , the body'surface temperature will depend on the outgoing flux $n_j q_j$

(2.5) $$T = T_e + \frac{1}{h} n_j q_j$$

If the convection exchange coefficient h tends to infinity, this reduces to a prescribed temperature condition. If it tends to zero, we have an adiabatic wall condition.

Should we write (2.5) in the equivalent form

$$n_j q_j = h(T - T_e) = q_e \tag{2.6}$$

the right hand side is an equivalent surface heat sink. We can then think of the rate of heat absorption q_e to be a prescribed quantity; in which case the bath temperature T_e must be adjusted to conform with the temperature reached by the body's surface.

The concept of a heat temperature can be extended, as a useful artifact, to interface transition conditions. On each interface A between two heat conducting bodies we can distinguish between two faces, conventionally denoted by A_+ and A_- (Fig. 1). The conditions generalizing (2.5) can then be written as

$$\begin{aligned} T_+ &= T_e + \frac{1}{h_+} n_{j+} q_{j+} \\ T_- &= T_e + \frac{1}{h_-} n_{j-} q_{j-} \end{aligned} \tag{2.5'}$$

If the interface temperature T_e is prescribed, each condition is independent of the other and there is, properly speaking, no transition condition. There is an interface heat sink

$$n_{j+} q_{j+} + n_{j-} q_{j-} = h_+(T_+ - T_e) + h_-(T_- - T_e) = q_e \tag{2.6'}$$

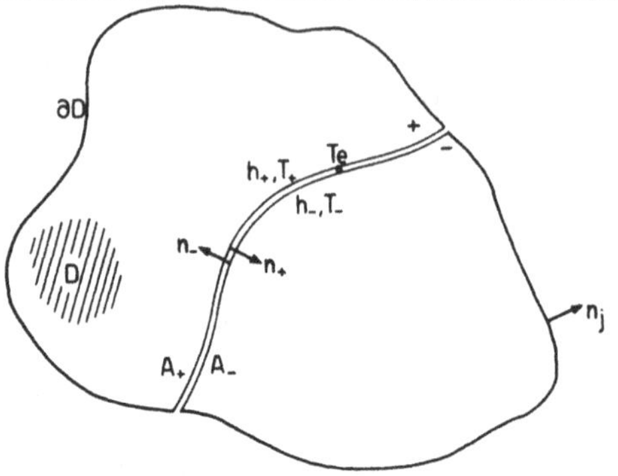

Fig. 1. Interface transition conditions

whose intensity has to be regulated to comply with the surface temperatures T_+ and T_-. Should we prescribe q_e instead of T_e, this last value must be adjusted to

$$T_e = \frac{h_+ T_+ + h_- T_- - q_e}{h_+ + h_-}$$

In this case, eliminating T_e between the two conditions (2.5'), the conditions can be written

$$T_+ - \frac{1}{h_+} n_{j+} q_+ = T_- - \frac{1}{h_-} n_{j-} q_-$$

(2.7)

$$n_{j+} q_{j+} + n_{j-} q_- = q_e \ .$$

Those are real transition conditions. Introducing an equivalent transitional convection coefficient h, defined by

(2.8) $$\frac{1}{h} = \frac{1}{h_+} + \frac{1}{h_-}$$

they can be put into the equivalent form

$$n_{j+} q_{j+} = h(T_+ - T_-) + \frac{h}{h_-} q_e$$

(2.7')

$$n_{j-} q_{j-} = h(T_- - T_+) + \frac{h}{h_+} q_e$$

Usually the prescribed heat sink value is zero; there is heat flow continuity and only the transitional convection coefficient h need be known.

If the interface is an artificial one, introduced for the purpose of numerical analysis by finite elements in a homogeneous body, we can let h_+ and h_- tend to infinity in the first of equations (2.7) and obtain the transitional requirement of temperature continuity.

3. VIRTUAL POWER AND HEAT SINK POTENTIAL

A vector field

$$e_i = \partial_i T \tag{3.1}$$

satisfies integrability conditions for the existence of T :

$$\partial_2 e_3 - \partial_3 e_2 = 0 \qquad \partial_3 e_1 - \partial_1 e_3 = 0 \qquad \partial_1 e_2 - \partial_2 e_1 = 0 \tag{3.2}$$

analogous to the compatibility equations a strain field must satisfy in order to integrate a displacement field.

Inasmuch as non-compatible strain fields do occur for certain types of approximation in elasticity theory, we are led to introduce the concept of "generalized temperature gradient field". Such a field is composed of e_i fields, that do not necessarily satisfy (3.1) in each subdomain, and of surface temperature discontinuities, either at interfaces or at the outer boundary; it will be symbolized by e .

Let q symbolize an independent heat flow field, piecewise differentiable in each subdomain. We then construct a "virtual power functional"

$$(e,q) = -\int_D e_i q_i \, dD + \int_S n_i q_i (T - T_e) \, dS \tag{3.3}$$

analogous to the structure of a virtual work in elasticity theory. S denotes the outer boundary plus the set of both faces of each interface and we recall that T_e has the same value on both faces of an interface.

If e is such that the vector field e obeys to (3.2) in each simply connected subdomain and a temperature field can be constructed, whose surface values coincide with those introduced in (3.3), it is called "compatible" and denoted by the symbol ∂T . Then, by simple integration by parts, we find

$$(\partial T, q) = \int_D T \, \partial_i q_i \, dD - \int_S n_i q_i T_e \, dS$$

Let us now denote by Q and q_e the heat sinks required to balance the flow field q , that is

$$\partial_i q_i = -Q \qquad \text{in} \quad D$$

$$n_i q_i = q_e \qquad \text{on} \quad \partial D$$

$$(n_i q_i)_+ + (n_i q_i)_- = q_e \qquad \text{on } A$$

We then obtain an identity between virtual power and a heat sink potential:

$$(\partial T, q) = -\int_{\partial D} T\, Q\, dD - \int_{A + \partial D} T_e\, q_e\, dS \tag{3.4}$$

In the heat sink potential, each heat sink intensity is multiplied by its corresponding temperature. This result is analogous to the identity between the internal form of virtual work and its external form (product of external forces by the conjugate displacements) when the strain field is compatible.

4. SCALAR PRODUCT AND DISSIPATION FUNCTIONAL

The result of section 3 makes no use of Fourier's law nor of the nature of the boundary conditions; it relies only on the concepts of compatibility for temperature gradients and balance, or equilibrium, for heat flow.

To any flow field we can now associate a generalized temperature gradient field, through Fourier's law

$$q_i = - k_{ij}\, e_j \tag{4.1}$$

and the heat convection equations

(4.2) $$n_i q_i = h(T - T_e)$$

applied on the totality of S . Substitution of this into (3.3) would create confusion by having the same symbol for the generalized temperature gradient field already present in (3.3) and the new field "conjugate" to q . For this reason we denote the latter one by $\hat{e}$, and obtain

(4.3) $$(e\,,\,\hat{e}) = \int_D k_{ij}\, e_i\, \hat{e}_j\, dD + \int_S h(T-T_e)(\hat{T}-T_e)\, dS$$

This expression is clearly in the nature of a scalar product between two generalized temperature fields; in particular

(4.4) $$(e\,,\,\hat{e}) = (\hat{e}\,,e\,)$$

Conversely, inverting Fourier's law

(4.5) $$e_j = -\, r_{jk}\, q_k$$

where r_{jk} is a heat resistivity tensor, reciprocal to the heat conductivity one,

(4.6) $$k_{ij}\; r_{jk} = \delta_{ik}$$

and using (4.2) in reverse, we are able to express the same scalar product in a dual form as

(4.7) $$(q\,,\hat{q}) = \int_D r_{ij}\, q_i\, \hat{q}_j\, dD + \int_S \frac{(n_i\, q_i)(n_j\, q_j)}{h}\, dS$$

In fact, since there is a one-one correspondance between a generalized temperature gradient field and its conjugate flow field, those fields represent the same element in function space. Thus, when we use the symbol f , it will represent either one of the conjugate fields. In this case (4.1) and (4.2) are analogous to the stress-strain relations.

5. THE DUAL VARIATIONAL PRINCIPLES

The definition of a scalar product induces a Hilbert space structure in the generalized temperature gradient fields (or their associated flow fields) provided

$$(f, f) = (e, e) = (q, q)$$

be positive definite. This is however a valid assumption, based on physical considerations related to the second law of thermodynamics. If Θ denotes the absolute value of the reference temperature, from which the (small) temperature excursion T are measured, then $(f, f)/\Theta^2$ is the (linearized) rate of increase of entropy of the body and its surroundings. We shall call

$$\frac{1}{2}(f, f) \quad \text{the dissipation functional}$$

It will prove convenient to use the property of a Hilbert space to be a metric space. If furnishes a concept of distance between two fields, the square of which is defined by

$$(f-\hat{f}, f-\hat{f}) \geq 0$$

and the positive definite property ensures that if the distance vanishes then $f = \hat{f}$.

Consider now a heat conduction problem where the volume heat sinks are specified

(5.1) $$Q = \bar{Q} \quad \text{in } D$$

and the boundary conditions are mixed. More precisely, we separate S in two subsets

$$S_1 \cup S_2 = S$$

and $A + \partial D$ in two corresponding subsets

$$B_1 \cup B_2 = A + \partial D$$

The set B_1 is the set S_1 where both faces belonging to the same interface have been merged. The heat sink temperatures are prescribed on B_1

(5.2) $$T_e = \bar{T}_e \quad \text{on } B_1$$

The set B_2 corresponds similarly to S_2 and the heat sink flow rate is specified on it

(5.3) $$q_e = \bar{q}_e \quad \text{on } B_2$$

Let ∂T_0 denote a fixed temperature gradient field with an integrated temperature distribution $T_0(x)$ in each subdomain and surface temperature discontinuities $T_0 - \bar{T}_e$ on B_1. Let $\partial \hat{T}$ denote an adjustable (containing unknown parameters) temperature gradient field, with integrated temperature distribution $\hat{T}(x)$ in each subdomain and surface temperature discontinuities $\hat{T} - 0$. Then, by the principle of superposition, $\partial(\hat{T} + T_0)$ will be an adjustable temperature gradient field, with integrated temperature distribution $\hat{T} + T_0$ and surface temperature discontinuities $\hat{T} + T_0 - \bar{T}_e$ on B_1.

Similarly introduce a fixed heat flow field, denoted by q_0 , which balances the heat sink rates $\bar{Q}$ in D and $\bar{q}_e$ on B_2. Introduce also an adjustable heat flow field $\hat{q}$ which balances zero heat sink rates in D and on B_2. Then, by superposition, $\hat{q} + q_0$ will be an adjustable heat flow field in equilibrium with the perscribed heat sink rates.

If the squared distance

$$(\partial\hat{T} + \partial T_0 - \hat{q} - q_0 \, , \, \partial\hat{T} + \partial T_0 - \hat{q} - q_0) \geq 0$$

happens to be zero, we have

$$\partial\hat{T} + \partial T_0 = \hat{q} + q_0$$

and both fields become the conjugate of one another and represent an exact solution to the problem. This solution is unique. Indeed if s_1 and s_2 are two solutions, $s_2 - s_1$ must be a temperature gradient field with $T_e = 0$ on B_1, that is of type $\partial\hat{T}$ Again $s_2 - s_1$ must also be a heat flow field in equilibrium with $\bar{Q} = 0$ in D and $\bar{q}_e = 0$ on B_2, that is of type $\hat{q}$. Thus

$$(s_2 - s_1 , s_2 - s_1) = (\partial\hat{T}, \hat{q}) = 0$$

in view of equation (3.4). This establishes that $s_2 - s_1$.

In numerical analysis the number of adjustable parameters is generally insufficient to produce an exact solution. A useful form of best approximation is then furnished by trying to minimize the squared distance between a temperature gradient formulation and a heat flow formulation:

$$(\partial\hat{T} + \partial T_0 - \hat{q} - \hat{q}_0 , \partial\hat{T} + \partial T_0 - \hat{q} - q_0) \text{ minimum}$$

In view of the result

$$(\partial\hat{T} , \hat{q}) = 0 \tag{5.4}$$

this is equivalent to

$$\begin{aligned}(\partial\hat{T} + \partial T_0 - q_0 , \partial\hat{T} + \partial T_0 - q_0) + (\hat{q} + q_0 - \partial T_0 , \hat{q} + q_0 - \partial T_0) \\ -(q_0 - \partial T_0 , q_0 - \partial T_0) \text{ minimum}\end{aligned} \tag{5.5}$$

The first term is non-negative and depends only on the adjustable parameters $\partial\hat{T}$, the second is also non-negative and depends only on the adjustable parameters in $\hat{q}$, the last one being fixed can be discarded. Consequently we may minimize separately the first and the second term, which procedure will furnish immediately the two most important variational principles governing steady heat conduction.

5.1 The principle of variation of temperatures

Minimization of the first term in (5.5) is equivalent to

$$\frac{1}{2}(\partial T, \partial T) - (\partial T, q_0) \quad \text{minimum} \qquad (T = \hat{T} + T_0) \quad (5.1.1)$$

Explicitly, using the definition of the scalar product and (3.4):

$$\frac{1}{2}\int_D k_{ij}\, \partial_i T\, \partial_j T\, dD + \frac{1}{2}\int_{S_1} h(T-\overline{T}_e)^2 dS + \frac{1}{2}\int_{S_2} h(T-T_e)^2\, dS$$
$$+\int_D T\, \overline{Q}\, dD + \int_{B_2} \overline{q}_e T_e dS + \int_{B_1} q_{e0} \overline{T}_e dS \quad \text{min.} \qquad (5.1.2)$$

The last term, that contains the fixed heat sink rates q_{e0} balancing the particular flow field q_0 on B_1 , can be discarded as constant. The Euler equations and boundary conditions of this principle of variation of temperatures are found to be

$$-\, \partial_i(k_{ij}\, \partial_j T) + \overline{Q} = 0 \qquad \text{in } D \qquad (5.1.3)$$

(5.1.4) $$n_i k_{ij} \partial_j T + h(T - \overline{T}_e) = 0 \qquad \text{on} \quad S_1$$

(5.1.5) $$n_i k_{ij} \partial_j T + h(T - T_e) = 0 \qquad \text{on} \quad S_2$$

stemming from volume and surface variations of the temperature field T. It must be understood that, according to whether parts of S_1 or S_2 belong to A_+ or A_-, h and T must be replaced by h_+ T_+ or h_-, T_- respectively.

An additional result is obtained from the arbitrary variation of the unspecified heat sink temperature T_e on S_2:

(5.1.6) $$-h_+(T_+ - T_e) - h_-(T_- - T_e) + \overline{q}_e = 0$$

or

(5.1.7) $$-h(T - T_e) + q_e = 0$$

according to whether $\overline{q}_e$ is at an interface or at the outer boundary. The correctness of equations (5.1.3) to (5.1.7) can be verified from the considerations developped in section 2.

An analysis of the limiting case, when a convection coefficient goes to infinity, shows that the corresponding contribution of the integrals on S_1 or S_2 in (5.1.2) must disappear; in compensation the corresponding face temperature must be set equal to $\overline{T}_e$ if it belongs to S_1, to T_e if it belongs to S_2.

5.2 The principle of variation of heat flows

Minimization of the second term in (5.5) is equivalent to

$$\frac{1}{2}(q\,,q) - (\partial T_0\,,q) \quad \text{minimum} \qquad (q = \hat{q} + q_0) \qquad (5.2.1)$$

Explicitly, and since for T_0 the heat sink temperature on B_1 is $\overline{T}_e$,

$$\frac{1}{2}\int_D r_{ij}\,q_i\,q_j\,dD + \frac{1}{2}\int_S \frac{(n_i q_i)^2}{h}\,dS + \int_D \overline{Q}\,T_0\,dD$$
$$+ \int_{B_1} q_e \overline{T}_e dS + \int_{B_2} \overline{q}_e T_{e0}\,dS \qquad \text{min.} \qquad (5.2.2)$$

The third and fifth terms, which are fixed constants, can be dropped.

In this "principle of variation of heat flow", analogous to the complementary energy principle of elasticity theory, the heat flow vector may not be varied unrestrictedly. It is constrained a priori to verify the heat balance equations:

$$\partial_j q_j + \overline{Q} = 0 \qquad \text{in} \quad D \qquad (5.2.3)$$

$$(n_j q_j)_+ + (n_j q_j)_- = \overline{q}_e \qquad \text{on} \quad B_2 \qquad (5.2.4)$$

or

$$n_j q_j = \bar{q}_e$$

Similarly, in the B_1 integral of (5.2.2) q_e must be understood to mean $(n_j q_j)_+ + (n_j q_j)_-$ or $n_j q_j$, according to whether the part of B_1 considered belongs to A or to ∂D.

This complementary variational principle can also be obtained from the temperature principle through a classical FRIEDRICHS transformation [5] . In so doing one also obtains useful two-field variational principles as intermediate steps. Those will not be considered here.

6. BOUNDING OF THE DISSIPATION FUNCTIONAL

To obtain a guaranteed upper or lower bound to the dissipation functional by approximations based on the preceeding variational principles, it is necessary to split the general heat conduction problem in two parts.

In part 1 we keep the given data $\bar{Q}$ in D and $\bar{q}_e$ on B_2 but set $\bar{T}_e = 0$ on B_1.

In part 2 we do the reverse, set $\bar{Q} = 0$ in D and $\bar{q}_e = 0$ on B_2 but introduce the data $\bar{T}_e$ on B_1.

By the principle of superposition, the general solutions is then obtained by adding the solution of parts 1 and 2.

6.1 Bounding of the dissipation functional in problem 1

Setting $\overline{T}_e = 0$ on B_1 means that we can dispense with the particular field T_0 and approximate the solution of problem 1 through either one of the variational principles

$$\frac{1}{2}(\partial\hat{T}, \partial\hat{T}) - (\partial\hat{T}, q_0) \quad \text{minimum} \qquad (6.1.1)$$

$$\frac{1}{2}(q, q) \quad \text{minimum} \qquad (6.1.2)$$

Take (6.1.1) first and assume a solution

$$\hat{T} = \alpha_i \hat{T}_{(i)} \qquad (6.1.3)$$

where the summation is finite and the $\hat{T}_{(i)}$ temperature fields have assumed distributions. Then (6.1.1) goes into the problem of minimizing a quadratic form in the unknown coefficients

$$\frac{1}{2}\alpha_i \alpha_j \left(\partial\hat{T}_{(i)}, \partial\hat{T}_{(j)}\right) - \alpha_i\left(\partial\hat{T}_{(i)}, q_0\right)$$

The coefficients are determined by solving the linear system

$$\alpha_j\left(\partial\hat{T}_{(i)}, \partial T_{(j)}\right) = \left(\partial\hat{T}_{(i)}, q_0\right) \qquad (i = 1, 2, \ldots, n)\ (6.1.4)$$

Denoting by $\hat{\Theta}$ the best approximate temperature field obtained by substituting those coefficients into (6.1.3), we see that (6.1.4) can be written

$$\left(\partial\hat{T}_{(i)} \, , \, \partial\hat{\Theta}\right) = \left(\partial\hat{T}_{(i)} \, , \, q_0\right) \qquad (i = 1,2, \ldots ,n)$$

Multipliyng each of these equations by the corresponding coefficient and adding

(6.1.5) $$(\partial\hat{\Theta}, \partial\hat{\Theta}) = (\partial\hat{\Theta}, q_0)$$

This property is analogous to the Clapeyron theorem of elasticity theory. It is obvious that it holds true also for the exact solution

(6.1.6) $$(s \, , s) = (s \, , q_0)$$

We recall that the symbol s for the exact solution, represents either the exact heat flow field or its conjugate temperature gradient field. Then the property

$$(s, s - q_0) = 0$$

equivalent to (6.1.6) follows directly from (5.4), because s as a temperature gradient field, is of type $\partial\hat{T}$ and $s - q_0$ as a flow field is of type $\hat{q}$ (balancing zero heat sink rates in D and on B_2). Since the approximate solution will not, in general, allow the minimum of (6.1.1) to be reached, we have

$$\frac{1}{2}(\partial\hat{\Theta} \, , \, \partial\hat{\Theta}) - (\partial\hat{\Theta}, q_0) \geq \frac{1}{2}(s \, , s) - (s, q_0)$$

Then, in view of (6.1.5) and (6.1.6), we obtain the boundings

$$(s,s) \geq (\partial\hat{\Theta}, \partial\hat{\Theta}) \qquad (s, q_0) \geq (\partial\hat{\Theta}, q_0) \tag{6.1.7}$$

Turn next to (6.1.2). Since again any best approximate solution p obtained from this variational principle, will not as a rule reach the absolute minimum:

$$(p, p) \geq (s, s) \tag{6.1.8}$$

Thus

$$(\partial\hat{\Theta}, \partial\hat{\Theta}) \leq (s, s) \leq (p, p) \tag{6.1.9}$$

This bracketing of the exact value of the dissipation functional through dual approximate analyses is very useful to appreciate the accuracy of the numerical solutions.

6.2 Bounding of the dissipation functional in problem 2

Setting $\bar{Q} = 0$ in D and $\bar{q}_e = 0$ on B_2 means that we can dispense of the particular flow field q_0 and approximate the solution of problem 2 trhough either one of the variational principles

$$\frac{1}{2}(\partial T, \partial T) \qquad \text{minimum} \tag{6.2.1}$$

$$\frac{1}{2}(\hat{q}, \hat{q}) - (\partial T_0, \hat{q}) \qquad \text{minimum} \tag{6.2.2}$$

Denoting by Θ the best approximate temperature field obtained through (6.2.1), it is clear that

$$(\partial\Theta , \partial\Theta) \geq (s , s)$$

Again, implementing (6.2.2) by a finite expansion

$$\hat{q} = \beta_i \hat{q}_{(i)}$$

we find that the best approximation $\hat{q} = \hat{p}$ has the property

(6.2.3) $$(\hat{p} , \hat{p}) = (\partial T_0 , \hat{p})$$

analogous to the Pasternak reduction theorem in structural analysis. This property is shared by the exact solution of problem 2

(6.2.4) $$(s , s) = (\partial T_0 , s)$$

because s , as a flow field, is of type $\hat{q}$ and $s - \partial T_0$, as a temperature gradient field, is of type $\partial \hat{T}$.

Furthermore

$$\frac{1}{2}(\hat{p} , \hat{p}) - (\partial T_0 , \hat{p}) \geq \frac{1}{2}(s , s) - (\partial T_0 , s)$$

Thus, in view of (6.2.3) and (6.2.4), we obtain

$$(s , s) \geq (\hat{p} , \hat{p}) \qquad (\partial T_0 , s) \geq (\partial T_0 , \hat{p})$$

The bracketing of the approximate analyses is now seen to be reversed

$$(\hat{p}, \hat{p}) \leq (s, s) \leq (\partial\Theta, \partial\Theta) \qquad (6.2.5)$$

which explains the necessity of splitting the general problem when it is wished to make use of this type of convergence criterion.

7. MATHEMATICAL MODELS OF TEMPERATURE FINITE ELEMENTS

These two minimum theorems may now be applied to the construction of mathematical models of finite elements either of the compatible temperature type or of the balanced heat flow type.

In this paper we shall restrain the discretization of the continuum problem to the most widely used way [6] i.e. discretization of the sole temperature field, continuous and piecewise differentiable.

The corresponding discretization of the temperature gradient follows through rigorous application of (2.1) and the whole weight of the approximation falls on the balance equation (2.4) which is only averaged as in a Galerkin process on each finite element domain by the shape functions chosen for the temperature field. The same is true for heat flux boundary specifications (5.3) which are translated into equivalent generalized fluxes and there are discontinuities in heat flow transmission between adjacent elements, the heat flux being soleley constrained [6] , [7] . On the other hand temperature boundary specifications (5.2) can be accounted for exactly if expressible in terms of the dito shaping functions.

The procedure is then to minimize the functional (5.1.2) (dissipation functional plus heat sink potential of prescribed thermal loads) with respect to the finite temperature degrees of freedom.

7.1 Temperature elements generation

Hereafter matrix formulation (*) will be used for brevity and simplicity. The temperature field is discretized by assuming

(7.1.1) $$T(x) = m^T(x)a = a^T m(x)$$

within the domain E of the finite element and on its boundary ∂E . The boundary consists of several parts $\partial_\sigma E(\sigma = 1, \dots, s)$ corresponding to adjacent elements or to parts of the external boundary ∂D .In (7.1) a and m are column vectors collecting respectively the parameters α_i to be determined by the minimum

---

(*) Unless otherwise specified, a small letter denotes a column vector, a capital letter denotes a matrix and superscript T denotes the transpose operation.

condition (5.1.2) (so that heat balance be approximated at best) and the assumed independent spatial modes $M_i(x)$ - usually polynomials - of the temperature field. Both the α_i - and M_i's are in finite number n.

Temperature gradients are readily

$$e = \partial T = D(x)\, a \tag{7.1.2}$$

where ∂ is the matrix differential operator

$$\partial^T = \left(\frac{\partial}{\partial x_1}, \frac{\partial}{\partial x_2}, \frac{\partial}{\partial x_3}\right)$$

and D a $(3 \times n)$ matrix of modes derivatives.

7.2 Local boundary temperatures and shaping functions

Continuity requirements for the temperature are still to be satisfied between elements. For this sake the behavior of $T(x)$ is analyzed on each face $\partial_\sigma E$ of the boundary ∂E : this furnishes the number of degrees of freedom of $T(x)$ on each $\partial_\sigma E$ and a local temperature coordinate is attached to each such degree of freedom. Local coordinates pertaining to $\partial_\sigma E$ are collected in the column matrix t_σ which determines completely $T(x)$ on it. Transition conditions are then satisfied along $\partial_\sigma E$ provided the same boundary modes exist for the adjacent element and the associated t_σ take the same value across the interface. Local temperatures here upon are related to the parameters of the field by

$$t_\sigma = R_\sigma a \tag{7.2.1}$$

Identification of temperature distribution along $\partial_\sigma E$ furnishes

$$m^T(x)\, a = p_\sigma^T(x)\, t_\sigma \qquad \text{for} \quad x \in \partial_\sigma E$$

where the elements of $p_\sigma(x)$ are the boundary temperature modes on $\partial_\sigma E$. This relation is valid for any a so that

$$m(x) = R_\sigma^T p_\sigma(x) \tag{7.2.2}$$

The set of relationships (7.2.1) for all the parts $\partial_\sigma E$ of the boundary results in the global matrix relation

$$t = R\, a \tag{7.2.3}$$

where t collects the n_b boundary local temperatures (some of them are readily common to several faces in order to insure uniqueness for temperature at vertices). This states that the knowledge of a determines completely the temperature on the boundary ∂E of the element but the question is now: does some value of a always correspond to an arbitrarily given t ?

The simplest case is that of a non singular matrix R implying that the number n of parameters be equal to the number n_b of boundary local temperatures. Inverting (7.2.3) and substituting into (7.1.1) there comes a translated discretization of the temperature field

$$T(x) = p^T(x)\, t \qquad (7.2.4)$$

where
$$p(x) = (R^T)^{-1} m(x) \qquad (7.2.5)$$

are the shaping functions of the temperature field. In other words the independent boundary temperatures are sufficient to determine the temperature within E . This is the case when the elements M_i of the spatial modes matrix $m(x)$ are the complete set of coordinates polynomials of degree $N \leq 2$, for finite elements of the $2D$- triangle of $3D$- tetrahedron type.

7.3 Local internal temperatures and bubble functions

The case where R is a non square matrix of rank n_b is still simple. In other words there are more field parameters α_i than local boundary temperatures $(n > n_b)$ and the way to by-pass this difficulty is to choice $(n - n_b = n_i)$ local internal temperatures ϑ to complete the set of n_b relations (7.2.3) by

$$\vartheta = S\, a \qquad (7.3.1)$$

where the row i of S corresponds to the row vector $m^T(x_i)$
The system

$$s = \begin{pmatrix} t \\ \vartheta \end{pmatrix} = \begin{pmatrix} R \\ S \end{pmatrix} a = N a \qquad (7.3.2)$$

can now be solved to give the general solution since the $(n \times n)$ matrix N is non singular. This furnishes

$$a = Ps = Ut + V\vartheta \tag{7.3.3}$$

and after substitution into (7.1) there comes

$$T(x) = p^T(x)\,t + q^T(x)\,\vartheta \tag{7.3.4}$$

where

$$p(x) = U^T m(x) \tag{7.3.5}$$

$$q(x) = V_m^T(x) \tag{7.3.6}$$

are respectively once more the shaping functions and the bubble functions. The latter represent indeed internal temperature fields which vanish on the boundary of the element.

7.4 Generalized heat fluxes

At this point all the required conditions but temperature boundary specifications are satisfied to invoque the minimum principle (5.1.2). In this appears the heat sink potential which introduces natural definitions for the generalized fluxes. The part due to internal heat generation gives in vue of (7.3.4)

$$\int_E \bar{Q}\,T(x)\,dE = t^T F_t + \vartheta^T F_\vartheta \tag{7.4.1}$$

where

$$f_t = \int_E \bar{Q}\, p(x)\, dE \qquad f_v = \int_E \bar{Q}\, q(x)\, dE \tag{7.4.2}$$

are defined as generalized heat fluxes conjugated respectively to the local temperature sets t and v. They appear as linear functionals where the actual prescribed heat generation is weighted by the shaping and bubble functions. Similarly for the (normal) heat flux at the boundary

$$\int_{\partial E} q_e T(x)\, d\partial E = t^T g_t \tag{7.4.3}$$

gives the definition of the generalized fluxes conjugated to the sole local boundary temperatures

$$g_t = \int_{\partial E} q_e\, p(x)\, d\partial E \tag{7.4.4}$$

The approximate solution to the temperature distribution using principle (5.1.2) will be shown depend only of the values of the generalized fluxes (7.4.2), (7.4.4) and many actual heat generation distributions and heat flux specifications can produce the same approximate answer. This is the announced characteristic feature of temperature finite elements models where the knowledge of the generalized heat fluxes corresponds to only a weak knowledge of the actual heat sink distribution.

7.5 Element conductivity matrix

Turning now to the dissipation term of principle (5.1.2) it can be computed directly for the element in terms of the parameters from the initial discretizaion (7.1.1) as

(7.5.1) $$F_i = \frac{1}{2}\int_E e^T C\, e\, dE = \frac{1}{2}\, a^T K_{aa}\, a$$

where F_i denotes the internal part of the dissipation functional and

$$K_{aa} = \int_E D^T(x)\, C\, D(x)\, dE \qquad C = (k_{ij})$$

and thereafter transformed through (7.3.3) in a quadratic form in the local temperatures

(7.5.2) $$F_i = \frac{1}{2}\, t^T K_t\, t + \upsilon^T K_{\upsilon t}\, t + \frac{1}{2}\, \upsilon^T K_{\upsilon\upsilon}\, \upsilon$$

with

$$K_{tt} = U^T K_{aa}\, U$$

$$K_{\upsilon t} = V^T K_{aa}\, U$$

$$K_{\upsilon\upsilon} = V^T K_{aa}\, V$$

It can also be obtained directly in the form (7.5.2) by performing integrations on the basis of the expansion (7.3.4) of T in shaping and bubble functions; in this case the generalized conductivity matrices are

$$K_{tt} = \int_E [\partial\, p^T(x)]^T \; C [\partial\, p^T(x)] dE$$

$$K_{vt} = \int_E [\partial\, q^T(x)]^T \; C [\partial\, p^T(x)] dE$$

$$K_{vv} = \int_E [\partial\, q^T(x)]^T \; C [\partial\, q^T(x)] dE$$

This last conductivity matrix is never singular because a bubble temperature function necessarily involves non zero temperature gradients within E ; this allows the associated local internal temperatures to be eliminated at the element level; indeed the minimization to be performed from principle (5.1.2) is finally in view of (7.4.1), (7.4.3) and (7.5.2) - i.e. without convection effect:

$$\frac{1}{2} t^T K_{tt} t + v^T K_{vt} t + \frac{1}{2} v^T K_{vv} v + t^T (f_t + g_t) + v^T f_v$$

minimum

The minimum conditions are

$$K_{tt} t + K_{tv} v = - f_t - g_t \qquad (7.5.3)$$

$$K_{vt} t + K_{vv} v = - f_v \qquad (7.5.4)$$

Solving the last one with respect to the internal temperatures pro duces

$$v = - K_{vv}^{-1} (f_v + K_{vt} t)$$

and the conductivity relation

$$K_E t_E = f_E + g_E \qquad (7.5.5)$$

with

$$K_E = K_{tt} - K_{tv} K_{vv}^{-1} K_{vt} \qquad (7.5.6)$$

$$f_E = - f_t + K_{tv} K_{vv}^{-1} f_v \qquad (7.5.7)$$

$$g_E = - g_t \qquad (7.5.8)$$

Relation (7.5.5) describes the discretized thermal properties of each element with respect to boundary temperatures $t_E = t$ but the case of the boundary part F_c of the dissipation functional need still be treated.

7.6 Specific convection boundary elements

This case is no longer treated by generalized fluxes [6] but results in essentially different elements as follows: an external degree of freedom is allocated to the boundary and a fictituous element is considered out of the actual body (Fig. 2). The local temperatures sequence for

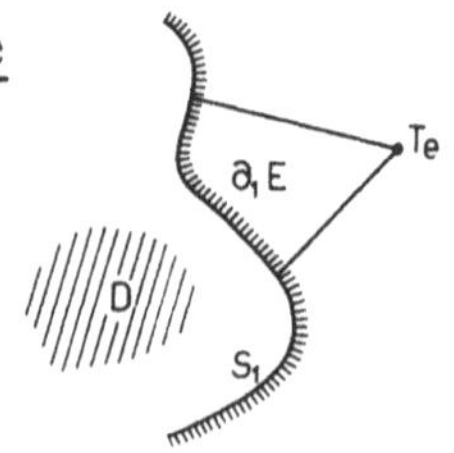

Fig. 2. Specific convection boundary element

this element comprises the set of temperatures pertaining to the boundary itself plus the external one T_e

$$t_c^T = (t^T T_e) \tag{7.6.1}$$

The external part F_c of the dissipation functional becomes then a quadratic form in these local temperatures

$$F_c = \frac{1}{2} \int_{\partial_1 E} h (T - T_e)^2 \, d\partial E - \frac{1}{2} t_c^T K_{cc} t_c \tag{7.6.2}$$

where $\partial_1 E$ is the part of the convection boundary S_1 allocated to the element. The minimization to be performed from principle (5.1.2) is in the present case

$$\frac{1}{2} t_c^T K_{cc} t_c + t_c^T g_c \quad \text{minimum}$$

where g_c is defined in a similar manner to (7.4.4) on $\partial_1 E$ The discretized thermal properties of those specific convection boundary elements are then summarized in

$$K_{cc} t_c = - g_c \tag{7.6.3}$$

7.7 Thermal properties of the assembled elements

The principle for assembling the elements is well known and fully described elsewhere [3] . Let us simply recall that it consists in stating which boundary temperatures must have common values at interfaces to implement the continuity requirements. The way to gather generalized fluxes follows from thermal balance at the interface. At this step it should be note that, while f_E and g_E (7.5.7), (7.6.8) are boundary generalized fluxes for the element, the g_E's are internal thermal loads for the whole body: this is clearly seen from the definition (7.4.2) of the f_E's where the generalized fluxes are computed from the known heat generation distribution while the boundary heat fluxes defining the g_E's by (7.4.4) are unknown.

Elementar conductivity matrices are also assembled in a master conductivity matrix and the thermal way of life of the whole body reads

$$K_s t_s = g_s \tag{7.7.1}$$

which can then be solved for the set of local temperatures t_s after due account is taken of the prescribed ones (boundary and external).

8. NUMERICAL EXAMPLES

The general method to derive temperature finite element models has been successfully applied [6] , [7] to several types of geometry (2D - triangle and quadrangle, AXISYM-triangle and 3D - tetraedron). The detailed matrices will not be presented here as long as the procedure to be followed is the standard one previously described.

Preference is given to some numerical illustrations which stress the efficiency of the method in solving steady state heat conduction problems.

8.1 Human body thermal model [8]

The thermal behavior of the human body has been modeled to simulate the major aspects of the heat transfer within the living tissues. The purpose is to develop thermoregulatory systems for protective suits used in current extravehicular activity space units. These systems must remove the metabolic heat generated by the body which was achieved in the most recent Apollo space suits by a network of water-cooled tubes in direct contact with the skin. The modeled human body is a set of rectangular strips (Fig. 3A) whose width is the half distance between cooling tubes. The depth is divided into three layers: an outer layer of skin, a skeletal muscle and a constant temperature inner core.

The tissues were considered to be isotropic and the thermal properties were taken as constant. The human body, in the vicinity of the parallel cooling tubes in contact with the skin, will be modeled as a parallelepiped with rectangular cross section. The thermal gradient parallel to the tubes will be considered zero so that the problem becomes two-dimensional

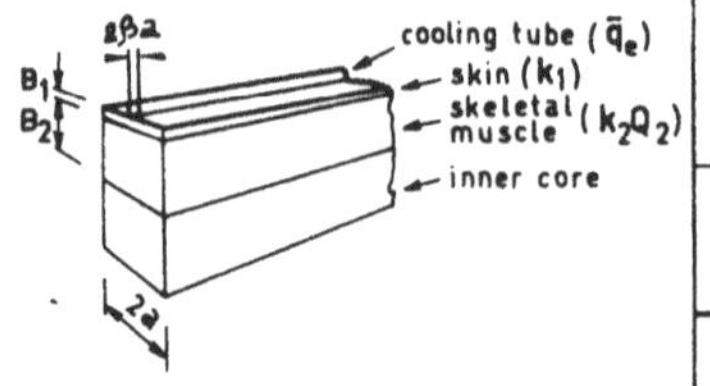

Fig. 3.a. Representative strip

CHARACTERISTIC DATA	
B_1 = 0,008	FT.
B_2 = 0,065	FT.
A = 0,032	FT.
K_1 = 0,242	BTU/HR-FT-°F
K_2 = 0,311	BTU/HR-FT-°F
Q_2 = 1398.	BTU/HR-FT^3
$\bar{q}_e$ = 1696.	BTU/HR-FT^2

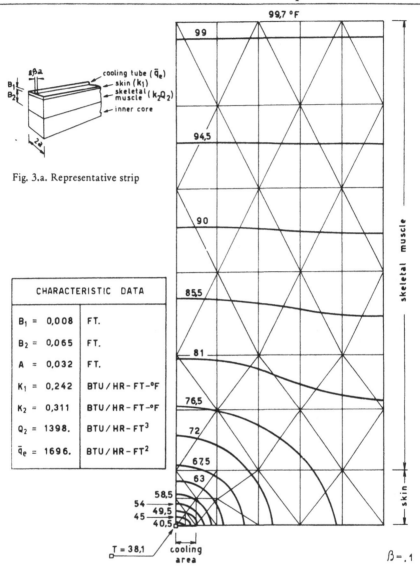

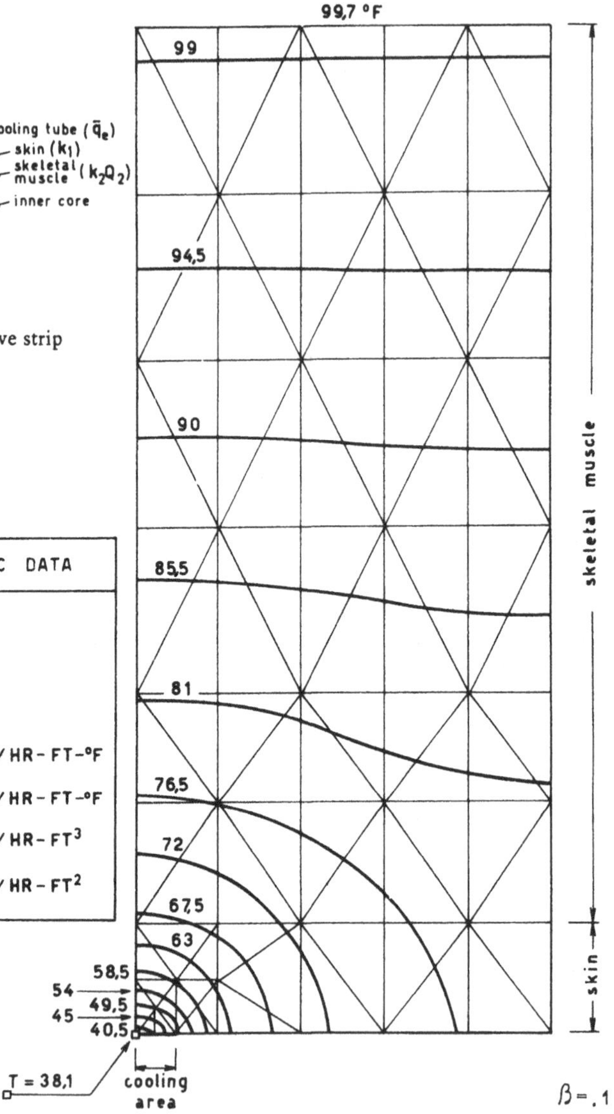

Fig. 3.b. Temperature distribution $\beta = .1$

in rectangular coordinates. The thermal field perpendicular to the tubes will be taken as symmetrical and the lines of symmetry which are located at the center of the cooling tubes and halfway between the tubes, are adiabatic boundaries. Since all the heat developed inside the body is assumed to be removed through the part β of the skin surface (prescribed heat flux), the remaining uncontacted part is also considered adiabatic. The inner boundary (interface core-skeletal muscle) is given the essentially constant temperature of the human body core (99,7° F).

Such simplifications are necessary to reach the limited data on the parameters involved and it is clear that the following numerical results are not immediately reliable for biological purposes.

Nevertheless we performed such a simplified numerical work through a 2D- finite element mesh of 82 parabolic T -field elements (Fig. 3.B). The temperature distribution is mapped for a case of high total metabolic rate (2600 Btu/hr = 760 watts). From a physiological viewpoint it is noteworthy that the required minimum skin temperature for the removal of 2600 Btu/hr is 38,1° F which is very low from a comfort standpoint.

8.2 Pressure vessel temperature distribution

As a second example, an axisymmetrical type of geometry is involved in computing the temperature distribution in the pressure vessel of a nuclear reactor (Fig. 4.A). The temperature map due to critical conditions is required for further thermal stresses analysis. Fig. 4.A presents an axial sec-

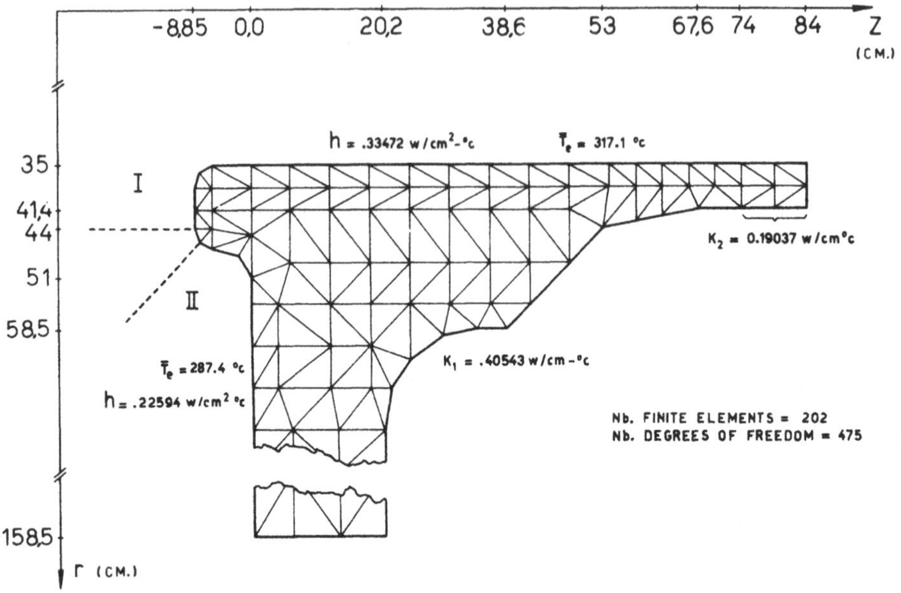

Fif. 4.a. Pressure vessel characteristics

tion of the hollow cylindrical type pressure vessel. The inner part noted I is occupied by a hot fluid and the corresponding boundary suffers convective heat transfer. The situation is analogous for part II of the boundary with a colder fluid. The remainder of the outer surface is considered adiabatic and the vessel isotropic and homogeneous, but for a bottom ring of different material.

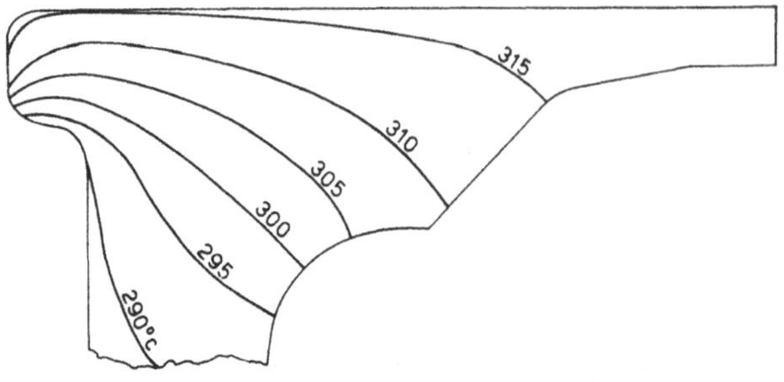

Fig. 4.b. Pressure vessel temperature distribution

Fig. 4.B presents the temperature map resulting from a treatment using a 202 finite element mesh of the parabolic temperature-type. In addition 38 specific convection boundary elements were needed to idealize the heat convection boundary.

REFERENCES

[1] B.M. Fraeijs de Veubeke, "Upper and lower bounds in matrix structural analysis", AGARDograph 72, Pergamon, 1964.

[2] B.M. Fraeijs de Veubeke, "Displacement and equilibrium models", Chpt 9. Stress Analysis, J. Wiley, 1965.

[3] G. Sander, "Application of the dual analysis principle", Proceedings of the IUTAM Symposium, Congrès et Colloques de l'Université de Liège, 1971.

[4] A.R.S. Ponter, "The application of dual minimum theorems to the finite element solution of potential problems with special reference to seepage", Int. Jnl for Num. Meth. in Eng., Vol. 4 , 1972.

[5] B.M. Fraeijs de Veubeke and M.A. Hogge, "Dual analysis for heat conduction problems by finite elements", Int. Jnl for Num. Meth. in Eng., Vol. 4, 1972.

[6] O.C. Zienkiewicz and Y.K. Cheung, Chp. 10, The Finite Element Method in Structural and Continuum Mechanics, McGraw Hill, London, 1967.

[7] M.A. Hogge, "Analyse numérique des problèmes thermiques en construction aéronautiqye et spatiale", Report SF-14, L.T.A.S., Université de Liège, 1970.

[8] J.C. Chato and A. Shitzer, "Thermal modeling of the human body - Further solutions of the steady-state equation", A.I.A.A. Journal, vol. 9, n° 5, pp. 865-869, May 1971.

PART II
UPPER BOUNDS TO A DISSIPATION FUNCTIONAL BY BALANCED HEAT FLUX MODELS

1. INTRODUCTION

A dual formulation of the boundary value problem of heat conduction was presented [1] , [2] and was shown to provide a numerical estimate of the accuracy of approximate solutions.

In this paper two ways of constructing mathematical models of finite elements based on the second single-field variational principle, i.e. principle of variation of heat flow, are described. The more direct way is to discretize balanced heat flow fields. Another way is to introduce a heat stream function, whereby the heat flow continuity at interfaces is obtained in a manner that is completely analogous to temperature continuity in temperature models. In both cases the knowledge obtained for the temperature field is weak since it will be based only on values of a finite number of functionals of that field. This does not appear to be a severe limitation since those weighted values are sufficient to give a realistic picture of the temperature distribution. Heat flow models enable numerical estimates to be made of the dissipation functional that bound the exact value from the opposite side to the estimates obtained from temperature models. This important evaluation of convergence is illustrates on several examples.

2. SUMMARY OF THE BASIC PRINCIPLE

The basic variational principle was obtained previously as

$$(2.1) \qquad \frac{1}{2}\int_D r_{ij} q_i q_j \, dD + \frac{1}{2}\int_S \frac{(n_i q_j)^2}{h} dS + \int_{B_1} q_e \bar{T}_e dS \quad \text{min}$$

$$(2.2) \qquad \left.\begin{array}{ll} \text{with} & q_e = (n_j q_j)_+ + (n_j q_j)_- \\ \text{or simply} & q_e = n_j q_j \end{array}\right\} \quad \text{on } B_1$$

We recall that B_1 is the set of interfaces or boundary surfaces where the heat sink temperature is prescribed. The heat flow vector in (2.1) must be constrained to satisfy "a priori"

$$(2.3) \qquad \partial_j q_j + \bar{Q} = 0 \qquad \text{in } D$$

$$(2.4) \qquad \left.\begin{array}{l} (n_j q_j)_+ + (n_j q_j)_- = \bar{q}_e \\ \text{or simply } n_j q_j = \bar{q}_e \end{array}\right\} \quad \text{on } B_2$$

where B_2 is the set of interfaces or boundary surfaces where the heat sink intensity is prescribed. Because of (2.3) and (2.4) the admissible heat flows are balanced heat flow fields.

Recuperation of the temperature field occurs in principle through the generalized temperature gradient field

$$e_j = - r_{jk} q_k \qquad \text{in } D \tag{2.5}$$

that can be obtained by inverting Fourier's law. However it does not in general comply with the integrability conditions

$$\partial_2 e_3 - \partial_3 e_2 = 0 \qquad \partial_3 e_1 - \partial_1 e_3 = 0 \qquad \partial_1 e_2 - \partial_2 e_1 = 0 \tag{2.6}$$

required to construct a pointwise defined temperature field T , such that

$$e_j = \partial_j T \tag{2.6}$$

The boundary conditions for T are

$$\left.\begin{array}{l} T_+ = \bar{T}_e + \left(\dfrac{n_j q_j}{h}\right)_+ \quad \text{and} \quad T_- = \bar{T}_e + \left(\dfrac{n_j q_j}{h}\right)_- \\ \text{or simply } T = \bar{T}_e + \dfrac{n_j q_j}{h} \end{array}\right\} \text{ on } B_1 \tag{2.8}$$

and similar conditions on B_2 , except that there T_e is not prescribed but unknown.

3. MATHEMATICAL MODELS OF HEAT FLOW FINITE ELEMENTS

3.1 Heat flow elements generation

Once again matrix formulation is used hence forward. The heat flow is discretized in the form

(3.1.1) $$q(x) = S(x)\,b + G(x)\,c$$

within the domain E of the finite element. b and c are column matrices of heat flow parameters to be determined by the minimum condition (2.1). The heat sink distribution in equilibrium with (3.1.1) must be

(3.1.2) $$\partial^T q(x) = \partial^T S(x)\,b + \partial^T G(x)\,c$$

where ∂^T is the matrix differential operator

$$\partial^T = \left(\frac{\partial}{\partial x_1} \quad \frac{\partial}{\partial x_2} \quad \frac{\partial}{\partial x_3}\right) .$$

The elements of $G(x)$ and $S(x)$ are usually taken to be polynomials of low degree. We have then distinguished those which generate heat flows in the absence of heat sinks, because

(3.1.3) $$\partial^T S(x) = 0$$

and those which correspond to certain modal distributions of heat sinks within the element

$$\bar{Q}(x) = \partial^T G(x) c = m^T(x) c \qquad (3.1.4)$$

3.2 Generalized heat fluxes

The boundary of the finite element consists of several parts $\partial_\sigma E(\sigma = 1,\ldots s)$. Assumption (3.1.1) generates along part $\partial_\sigma E$ of the boundary a set of independent heat flux modes to each of which a generalized heat flux coordinate is attached. The set of generalized flux coordinates for $\partial_\sigma E$ constitutes a column matrix g_σ and the following identity must hold for arbitrary b and c

$$n_\sigma^T S(x) b + n_\sigma^T G(x) c = f_\sigma^T(x) g_\sigma \qquad x \in \partial_\sigma E$$

where

$$f_\sigma^T = \left[f_1(x) \ldots f_{k_\sigma}(x) \right] \qquad x \in \partial_\sigma E \qquad (3.2.1)$$

$$n_\sigma^T = \left[n_1(x) \quad n_2(x) \quad n_3(x) \right] \qquad x \in \partial_\sigma E$$

are respectively the set of k_σ independent heat flux modes and the outward normal components along $\partial_\sigma E$.

In fact definition (3.2.1) of a generalized heat flux as intensity factor of a heat flux mode produces its dependence on the values of the parameters b and c ; in other words we are provided with a relationship of the form

$$g_\sigma = A_\sigma b + B_\sigma c \tag{3.2.2}$$

and identity (3.2.1) corresponds to

$$n_\sigma^T S(x) = f_\sigma^T A_\sigma \qquad n_\sigma^T G(x) = f_\sigma^T B_\sigma \qquad x \in \partial_\sigma E \tag{3.2.3}$$

The knowledge of g_σ determines the complete heat flux distribution along $\partial_\sigma E$ so that continuity of heat fluxes at interfaces can be secured simply by reciprocity of corresponding generalized flux coordinates.

Any predetermined set of boundary heat flux modes may be replaced by a new set of k_σ linearly independent combinations of the elements of the old set. Advantage is taken of this freedom to give simple physical significance to the generalized flux coordinates, either as local heat flux values or moments of various order. The set of relations (3.2.2) for all parts of the boundary ∂E results in a global relation

$$g_b = A b + B c \tag{3.2.4}$$

where the boundary flux coordinates g_σ are listed in g_b in some conventional order and A and B correspondingly partitioned from the $(k_\sigma \times n)$ and $(k_\sigma \times t)$ matrices A_σ and B_σ

3.3 Generalized temperatures

The heat sink potential part of functional (2.1) is next calculated by assuming the external temperature T_e to be defined along the whole boundary ∂E ; it then becomes formally

$$P = \sum_{\sigma} \int_{\partial_{\sigma} E} T_e n_{\sigma}^{T} q(x)\, d\partial E \qquad x \in \partial_{\sigma} E \qquad (3.3.1)$$
$$= \sum_{\sigma} g_{\sigma}^{T} \int_{\partial_{\sigma} E} T_e f_{\sigma}(x)\, d\partial E$$

We now define generalized temperature coordinates t_{σ} along each boundary by

$$P = \sum_{\sigma} g_{\sigma}^{T} t_{\sigma} \qquad x \in \partial_{\sigma} E \qquad (3.3.2)$$

and consequently the generalized temperatures sequence along $\partial_{\sigma} E$ collects, in the same order as in g_{σ} , linear functionals of the unknown temperature such as

$$t_j = \int_{\partial_{\sigma} E} T_e f_j(x)\, d\partial E \qquad x \in \partial_{\sigma} E \qquad j = 1, \ldots k_{\sigma} \qquad (3.3.3)$$

where the boundary heat flux modes play the role of weighting functions. Again if the coordinates t_{σ} of each boundary are collected in a single colums matrix t_b , in the same order as the g_σ's in g_b , the heat sink potential (3.3.2) takes the form

(3.3.4) $$P = g_b^T t_b$$

which summarizes in a concise way the conjugate character of generalized fluxes and generalized temperatures.

Besides, if internal heat sink modes (3.1.4) are considered, conjugate generalized temperatures can be defined by analyzing the remaining heat sink potential

(3.3.5) $$\int_E T\, Q\, dE = \left[\int_E T\, m^T(x)\, dE\right] c = t_i^T c$$

To each internal heat sink coordinate in c a conjugate generalized temperature in t_i is thus attached, such that

(3.3.6) $$t_i = \int_E T\, m(x)\, dE$$

Our knowledge of the temperature field will remain the "weak" one provided by the values taken by the linear functionals (3.3.3) and (3.3.6). This is expected since the discretized temperature gradient field (2.5)

$$e(x) = -\,C^{-1} q(x) \qquad C^{-1} = (r_{ij})$$

will in general not be integrable.

3.4 Generalized conductivity relations

As suggested hereupon we wish to extend the principle of variations of heat flow to the situation where internal heat sink distribution instead of being fixed, is variable and depends on adjustable parameters c .

Their variations are then constrained by the heat balance equation

$$\partial^T \delta q(x) + m^T(x)\, \delta c = 0 \tag{3.4.1}$$

This condition is removed by a Lagrangian multiplier λ and added to the variation of the internal part of the dissipation functional

$$\delta F_i + \int_E \left[e^T\, \delta q(x) + \lambda\, \partial^T \delta q(x) + \lambda\, m^T(x) \delta c \right] dE = 0$$

Partial integration on the second term gives

$$\delta F_i + \int_E \left[(e^T - \delta^T \lambda) \delta q(x) + \lambda\, m^T(x) \delta c \right] dE$$

$$+ \int_{\partial E} \lambda\, n^T\, \delta q(x)\, d\partial E = 0$$

and introduces in the principle of variation of heat flow the additional term

(3.4.2) $$\left[\int_E T\, m^T(x)\, dE\right] \delta c = t_i^T \delta c$$

if λ is taken as the integrated gradient field **e** and $\partial E \in S_1$. This is now applied to discretization (3.1.1). The internal dissipation functional is a quadratic positive form

(3.4.3) $$F_i = \frac{1}{2}\int_E q^T(x) C^{-1} q(x)\, dE = \frac{1}{2} b^T R^1_{bb} b + b^T R^1_{bc} c + \frac{1}{2} c^T R^1_{bc} c$$

with generalized resistivity matrices

$$R^1_{bb} = \int_E S^T C^{-1} S\, dE$$

$$R^1_{bc} = \int_E S^T C^{-1} G\, dE$$

(3.4.4) $$R^1_{cc} = \int_E G^T C^{-1} G\, dE$$

The boundary part of the dissipation functional leads also to a quadratic form through the heat flux

(3.4.5) $$n_c^T(x)\, q(x) = b^T A_\sigma^T f_\sigma + c^T B_\sigma^T f_\sigma$$

as

$$F_b = \frac{1}{2} \sum_{\sigma} \frac{1}{h} \int_{\partial_\sigma E} (n_\sigma^T q)^2 \, d\partial E$$

$$= \frac{1}{2} b^T R_{bb}^2 b + b^T R_{bc}^2 c + \frac{1}{2} c^T R_{cc}^2 c \qquad (3.6)$$

with additional generalized resistivity matrices

$$R_{bb}^2 = \sum_{\sigma} A_\sigma^T H_\sigma A_\sigma$$

$$R_{bc}^2 = \sum_{\sigma} A_\sigma^T H_\sigma B_\sigma \qquad (3.7)$$

$$R_{cc}^2 = \sum_{\sigma} B_\sigma^T H_\sigma B_\sigma$$

where $H_\sigma = \frac{1}{h} \int_{\partial_\sigma E} f_\sigma f_\sigma^T \, d\partial E$ (3.8)

We recall here that advantage is taken of the subdivision of the structure in elements to incorporate eventual heat convection losses at interfaces and not only along the external boundary ∂D.

The global dissipation functional reads now

$$\frac{1}{2} b^T R_{bb} b + b^T R_{bc} c + \frac{1}{2} c^T R_{cc} c \qquad (3.9)$$

if the resistivity matrices collect the separate ones as for instance

$$R_{bb} = \overset{1}{R}_{bb} + \overset{2}{R}_{bb} \tag{3.4.10}$$

The global heat sink potential yields from (3.3.4) and (3.3.5)

$$t_b^T g_b + t_i^T c = t_b^T A\, b + \left(t_b^T B + t_i^T\right) c$$

Substitution of this and (3.4.9) into principle of variation of heat flow and identification of the coefficients of the variation δb and δc produces

$$\begin{aligned} R_{bb} b + R_{bc} c &= - A^T t_b \\ R_{cb} b + R_{cc} c &= - B^T t_b - t_i \end{aligned} \tag{3.4.11}$$

The first equation may be solved for b and substituted into relation (3.2.4) as long as the parameters b are independent and each of them represent a non zero contribution to the dissipation functional; this furnishes

$$g_b + \left(A\, R_{bb}^{-1} R_{bc} - B\right) c = - A R_{bb}^{-1} A^T t_b \tag{3.4.12}$$

which is a conductivity relation completely analogous to that of a temperature-type element. The internal heat sink distribution is supposedly given, i.e. c is known, and the way it acts on the discretized behaviour of the element is through internal generalized fluxes given by

$$g_i = \left(A R_{bb}^{-1} R_{bc} - B\right) c \qquad (3.4.13)$$

The general conductivity matrix

$$K_b = - A R_{bb}^{-1} A^T \qquad (3.4.14)$$

involves only part A of the heat flux connection matrix. If an estimation of the generalized temperatures t_i conjugated to the generalized internal heat fluxes is desired, we invoke the second relation of (3.4.11) to produce

$$t_i = \left(R_{cb} R_{bb}^{-1} A^T - B^T\right) t_b + \left(R_{cb} R_{bb}^{-1} R_{bc} - R_{cc}\right) c \qquad (3.4.15)$$

In summary the discretized thermal relations of the element are

$$K_b t_b = g_b + g_i \qquad (3.4.16)$$

The way to collect elements together is then the same as for the temperature models [1] .

3.5 Remarks on heat flow finite elements

Generalized conductivity matrices are always singular because the homogeneous system

$$K_b t_b = 0 \qquad (3.5.1)$$

has amongst its non trivial solutions at least the uniform reference temperature of the element. These solutions are identical to those of

$$A^T t_b = 0 \tag{3.5.2}$$

since R_{bb}^{-1} is positive definite, and let us show that an uniform temperature field T_0 is solution of (3.5.2). Indeed the corresponding boundary and internal generalized temperatures are obtained directly by inserting this field into the definitions (3.3.3) and (3.3.6). Then, in view of the balance equation satisfied by any heat flow field

$$\partial^T q(x) + m^T(x) c = 0 \tag{3.5.3}$$

integrating over the finite element domain E with T_0 as weighting function, we obtain

$$-\int_E q^T(x)\, \partial T_0 \, dE + \int_{\partial E} T_0 n^T(x) q(x) \, d\partial E + \left\{ \int_E T_0 m^T(x) \, dE \right\} c = 0$$

where an integration by parts has been performed on the first term of (3.5.3). Now the reference temperature field produces no temperature gradient field and expressions (3.3.4) and (3.3.5) of the associated heat sink potential furnish

$$g_b^{T} t_b^{o} + c^{T} t_i^{o} = 0 \tag{3.5.4}$$

which expresses the vanishing of the total outgoing heat flux and is only normal because of the detailed equilibrium equations (2.3) and (2.4): it is the analogue of the overall equilibrium conditions between generalized forces in elasticity.

Substituting decomposition (3.2.4) of the g_b's and noting that the relation holds for arbitrary b and c, there follows

$$A^{T} t_b^{o} = 0 \tag{3.5.5}$$

$$B^{T} t_b^{o} + t_i^{o} = 0 \tag{3.5.6}$$

This completes the proof for the uniform reference temperature field. Usually there are no other non zero solutions to (3.5.5) and the element is a "sound" one. If the case may be, an internal temperature field is taken from (3.5.6)

$$t_i = - B^{T} t_b$$

and relations (3.4.11) lead to the vanishing of b and c.

In other words a temperature field exists that produces no heat flow. This is however not the case for the family of 2-D triangles or 3-D tetrahedra elements provided the approximate functions in (3.1.2) are a complete set of polynomials.

4. EXAMPLES AND NUMERICAL EFFECTIVENESS OF HEAT FLOW ELEMENTS

The general method to derive heat flow finite elements has been successfully applied [2] , [3] to several degrees of polynomial approximation for the heat flow. The simplest 2-D elements will be briefly described and attention focused on an example with internal heat sink distribution. A numerical example shows that the generalized temperatures associated with heat flow models give a good idea of local distribution even in case of non integrable gradient fields.

4.1 Constant heat flow model (fig. 1A)

The simplest case when no internal heat sinks are considered and when the polynomial degree for the heat flow is the lowest is very easy to handle. Heat flux at interface is constant and a single generalized heat flux is sufficient to insure perfect transmission to the adjacent element. Consequently there is only one generalized temperature by interface, which is the mean of the temperature distribution a-

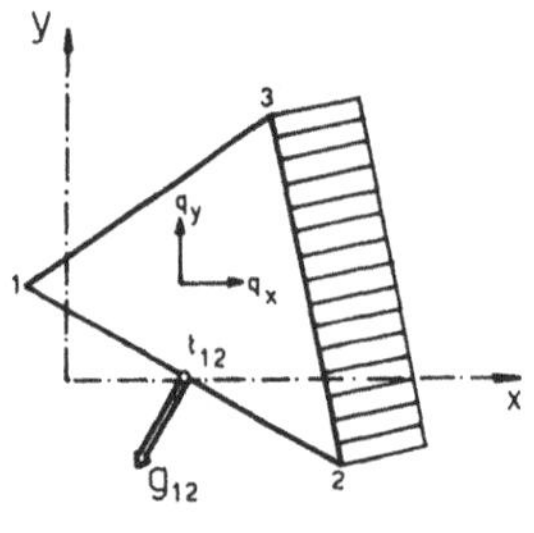

$$q_x = \alpha_1$$
$$q_y = \alpha_2$$
$$t_{ij} = \frac{1}{L_{ij}} \int_{ij} T_e \, ds$$

Fig. 1.a. Constant heat flow model

long the interface and is located at mid-edge.

4.2 Linear heat flow model (Fig. 1B)

The heat flow components are written in a form implementing "a priori" the homogeneous heat balance equation and depend so far on five independent parameters. This time the heat flux at interfaces is linearly varying and requests two two generalized coordinates to enable identification of the whole heat flux between edjacent elements. The conjugate generalized temperatures are once more weighted averages of the temperature distribution along the interface located to each third of the edge.

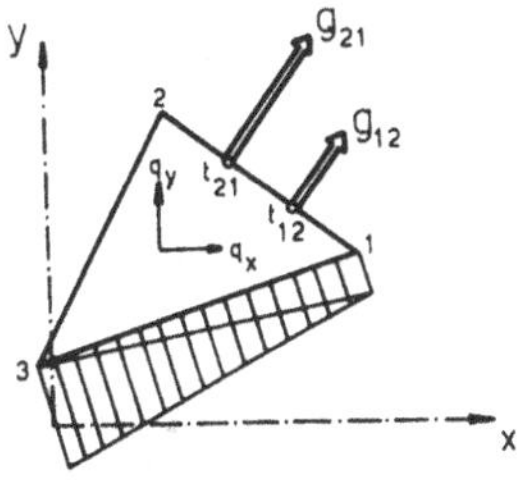

$$q_x = \alpha_1 + \alpha_2 X + \alpha_3 Y$$

$$q_y = \alpha_4 + \alpha_5 X - \alpha_2 Y$$

$$t_{ij} = \frac{2}{L^2_{ij}} \int_{ij} T_e (L_{ij} - s)\, ds$$

$$t_{ji} = \frac{2}{L^2_{ij}} \int_{ij} T_e\, s\, ds$$

Fig. 1 b. Linear heat flow model

4.3 Heat flow model with internal heat sinks

We will derive now a general but still simple model according to the standard procedure developed in section 3. It is a 2D- triangle (Fig. 1.C) similar to the preceding

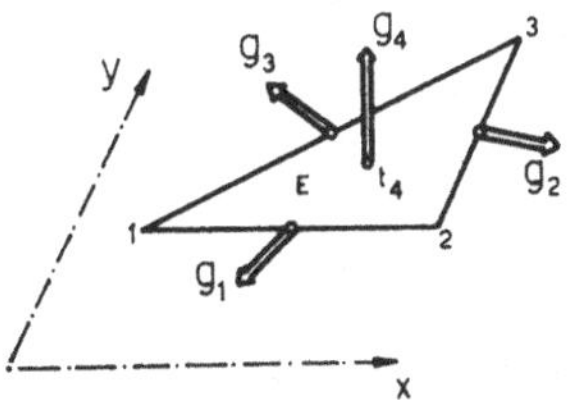

$$t_4 = \int_E T\, dE$$

Fig. 1.c. Constant heat flow model with internal heat generation ones but we want to take into account a possible internal heat sink distribution. The following discretization is assumed

$$(4.3.1)\qquad q_x = \alpha_1 - \frac{Q}{2} x \qquad q_y = \alpha_2 - \frac{Q}{2} y$$

which can be identified to (3.1.1) provided

$$(4.3.2)\qquad q^T = (q_x q_y)\,; \quad b^T = (\alpha_1\ \alpha_2)\,; \quad c = -\frac{Q}{2} \qquad S = E\,; \quad G^T = (x\ y)\,; \quad m^T = (1\ \ 1)$$

The generalized fluxes are next calculated; assumption (4.3.1) generates only a constant heat flux distribution along the sides of the triangle as can be easily verified from (3.2.1); in this case $k_\sigma = 1$ and $f_\sigma = \frac{1}{L_{ij}}$ and relation (3.2.4) becomes

$$\begin{vmatrix} g_1 \\ g_2 \\ g_3 \end{vmatrix} = \begin{vmatrix} y_{21} & x_{12} \\ y_{32} & x_{23} \\ y_{13} & x_{31} \end{vmatrix} b - \begin{vmatrix} x_1 y_{21} + y_1 x_{12} \\ x_2 y_{32} + y_2 x_{23} \\ x_3 y_{13} + y_3 x_{31} \end{vmatrix} \frac{Q}{2}$$

where $x_{ij} = x_i - x_j$ $\quad y_{ij} = y_i - y_j$ $\quad i,j =$ vertices of the triangle

The conjugate generalized boundary temperatures are then the means

$$t_1 = \frac{1}{L_{12}} \int_{\partial_\sigma E=12} T_e \, d\partial E\,; \qquad t = \frac{1}{L_{23}} \int_{\partial_\sigma E=23} T_e \, d\partial E\,;$$

$$t_3 = \frac{1}{L_{31}} \int_{\partial_\sigma E=31} T_e \, d\partial E$$

as can be achieved from (3.3.3) and are identical to those of the element described in 4.1. The additional heat sink coordinate c introduces a fourth generalized temperature which is following (3.3.6)

$$t_4 = \int_E T \, dE$$

We possess at this point all the elements to perform the detailed resistivity matrices of the element as from (3.4.4) and definitions (4.3.2)

$$R^1_{bb} = E \begin{vmatrix} \frac{1}{k_x} & 0 \\ 0 & \frac{1}{k_y} \end{vmatrix}\,; \quad R^1_{bc} = \begin{vmatrix} \frac{S_x}{k_x} \\ \frac{S_y}{k_y} \end{vmatrix}\,; \quad R^1_{cc} = \left(\frac{I_x}{k_x} + \frac{I_y}{k_y} \right)$$

where k_x, k_y are the orthotropic conductivity coefficients and

$$E = \int_E dE \qquad S_x = \int_E x\, dE \qquad S_y = \int_E y\, dE$$

$$I_x = \int_E x^2\, dE \qquad I_y = \int_E y^2\, dE$$

The conductivity relation is then detailed following (3.4.12), (3.4.13), (3.4.14).

4.4 Numerical example

Figure 2.A illustrates a conduction-cooled turbine blade having its root maintained at a uniform temperature

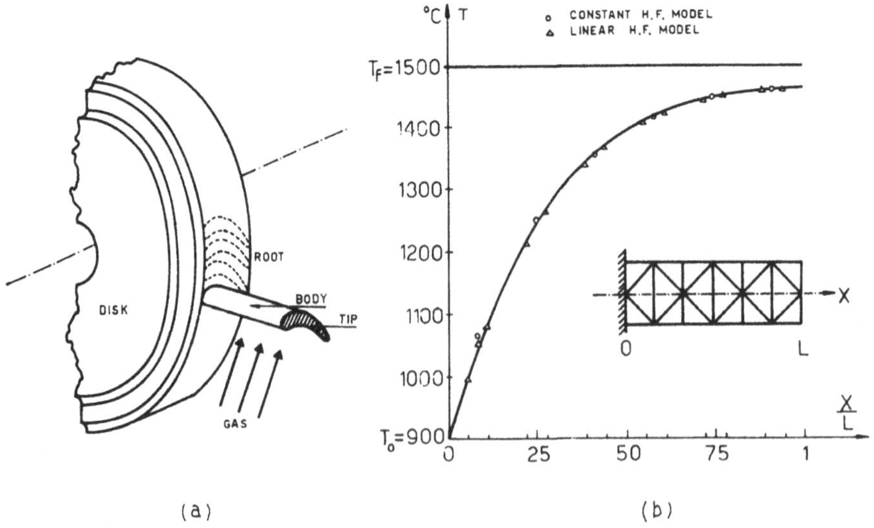

Fig. 2. Conduction cooled turbine blade analysis by heat flow elements

T_0 and operating in an effective gas with temperature T_f. The tip is considered to be insulated, the body exchanges heat with the gas by convection and the disk acts as a heat sink to cool the blade. A finite element analysis (Fig. 2.B) has been performed on the basis of a 24 triangular mesh of successively constant and linear heat flow models. The temperature plotting shows a remarkable accuracy for the corresponding generalized temperatures versus the unidimensional analytical solution. This attempts to demonstrate that the abstract concept of generalized temperature gives a good picture of the detailed local field even in case of non integrable gradient fields, provided some physical sense guides the choice and interpretation of those generalized quantities.

5. CONSTRUCTION OF HEAT FLOW MODELS BY MEANS OF HEAT STREAM FUNCTIONS

Recourse to the concept of weak generalized temperatures to assemble heat flow models may be responsible for some lack of understanding in the real usefulness and value of the model. As a matter of fact, if informations on temperature distributions are desired, it is the most natural way to develop the heat flow element. However, if only heat flow information is requested, temperature concepts can seemingly be discarded by using heat stream functions.

The introduction of heat stream functions allows

automatic satisfaction of the heat flow balance in the absence of volume heat sinks and this will be assumed to be the nature of the problem. Two cases will be considered:

- two-dimensional flow where a scalar stream function is sufficient to express balance;
- three-dimensional flow where a vector-type of stream function is necessary.

5.1 Two-dimensional heat flow models based on a stream function

The homogeneous form of the balance equation

(5.1.1) $$\partial_j q_j = 0$$

is satisfied by using a stream function ψ such that

(5.1.2) $$q_1 = \partial_2 \psi \qquad\qquad q_2 = - \partial_1 \psi$$

or

(5.1.3) $$q_j = \varepsilon_{jm} \partial_m \psi$$

where ε_{jm} is the two-dimensional alternator symbol .

The flux through a boundary of element described in anti-clockwise sense (Fig. 3.A) is then given by

(5.1.4) $$n_j q_j = \varepsilon_{jm} n_j \partial_m \psi = n_1 \partial_2 \psi - n_2 \partial_1 \psi = \frac{\partial \psi}{\partial s}$$

We know that in the heat flow variational principle, q_j must not only satisfy the homogeneous balance equation (5.1.1), where we assumed restriction to $\bar{Q}=0$, but also

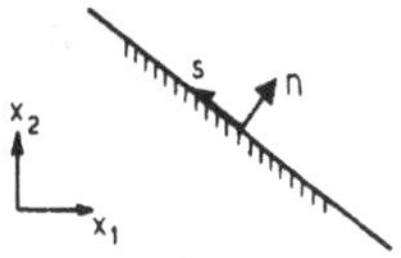

Fig. 3.a.

$$(n_j q)_+ + (n_j q_j)_- = \bar{q}_e$$

on B_2

or

$$n_j q_j = \bar{q}_e$$

For simplicity we take $\bar{q}_e = 0$ at all interfaces, consequently discarding all artificial impositions of temperatures $\bar{T}_e$ at interfaces. Thus we consider only as boundary and interface conditions

$$(n_j q_j)_+ + (n_j q_j)_- = 0 \quad \text{on } A \quad \text{set of all interfaces} \qquad (5.1.5)$$

$$n_j q_j = \bar{q}_e \qquad \text{on } \partial_2 D \qquad (5.1.6)$$

$$T_e = \bar{T}_e \qquad \text{on } \partial_1 D \qquad (5.1.7)$$

Considering (5.1.4), (5.1.5) can be rewritten in the form

$$\left(\frac{\partial \psi}{\partial s}\right)_+ + \left(\frac{\partial \psi}{\partial s}\right)_- = 0 \quad \text{on } A \qquad (5.1.8)$$

Noting that ds_+ and ds_- are described in opposite senses, this may be satisfied by simple continuity of the stream function ψ across the interfaces. The condition, if sufficient, is however not quite necessary since (5.1.8) obviously allows a constant difference between the two face values of ψ. This jump does in fact occur for multiply-connected domains D, as is well known and will be briefly recalled.

Take first a simply connected domain, like that of each finite element itself. Then we have everywhere inside satisfaction of (5.1.1) and

$$\int_D \partial_j q_j \, dD = 0$$

This is, by partial integration, equivalent to

$$\int_{\partial D} n_j q_j \, ds = \int_{\partial D} \frac{\partial \psi}{\partial s} ds = \int_{\partial D} d\psi = 0$$

Thus ψ is single valued throughout the domain. Consider now a domain with an internal cavity (Fig. 3.B) of contour C_0. Here

$$\int_{C_0} d\psi = \int_{C_0} n_j q_j \, ds = g_0 \tag{5.1.9}$$

where g_0 is the total heat flux penetrating into the cavity, a quantity that needs not be necessarily zero. Thus ψ has a cyclic constant when the contour of the cavity is described. The same cyclic constant is valid for any closed contour C, described in the domain and circling once around this cavity. For, if we join c_0 and c by a two-faced cut 1234, the integral

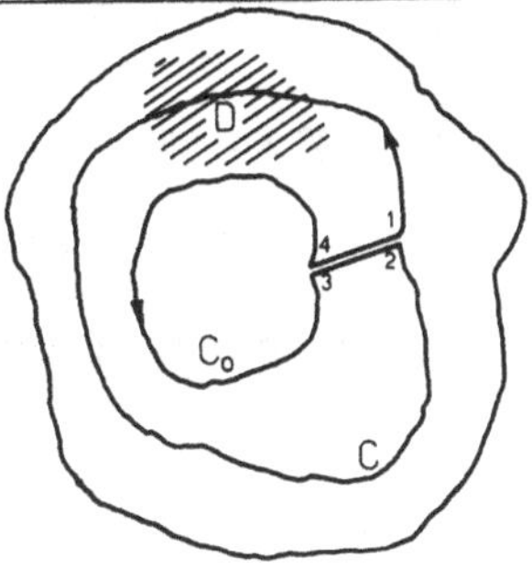

Fig. 3.b.

$$\int_{12341} d\psi = \int_{C} d\psi + \int_{2}^{3} d\psi - \int_{C_0} d\psi + \int_{4}^{1} d\psi = 0$$

because the inner domain of the contour 12341 is simply connected. However, as there is reciprocity of heat flux across the cut, the second and last integrals cancel, leaving

$$\int_{C} d\psi = \int_{C_0} d\psi = g_0 \qquad (5.1.10)$$

For this reason it will be necessary in the discretization process by finite elements to isolate a sequence of interfaces providing a barrier extending from the inner cavity to the external contour, across which a constant jump in the value of ψ must be left open.

The generalization to several internal cavities is immediate.

In terms of the heat stream function the heat flow variational principle (2.1) reads

$$\frac{1}{2}\int_{D} r_{ij}\, \epsilon_{im}\, \epsilon_{jn}\, \partial_m \psi\, \partial_n \psi\, dD + \frac{1}{2}\int_{S} \frac{1}{h}\left(\frac{\partial\psi}{\partial s}\right)^2 ds + \int_{\partial_1 D} \frac{\partial\psi}{\partial s}\, \overline{T}_e\, ds \text{ minimum}$$

(5.1.11)

The continuity of ψ holds on all interfaces except the barriers $b_1 \ldots b_m$ whose separation make D simply connected. The only "a priori" conditions remaining on ψ is (5.1.6) in its new form

$$\frac{\partial \psi}{\partial s} = \bar{q}_e \qquad \text{on } \partial_2 D \tag{5.1.12}$$

Before discretizing let us examine the Euler equations and natural boundary conditions of this new principle.

Variation of the first integral

$$\frac{1}{2}\int_D r_{ij}\, \varepsilon_{im}\, \varepsilon_{jn}(\partial_m \psi\, \partial_n\, \delta\psi + \partial_n \psi\, \partial_m\, \delta\psi)\, dD$$

is first reduced to

$$\int_D (r_{ij}\, \varepsilon_{im}\, \varepsilon_{jn}\, \partial_n \psi\, \partial_m\, \delta\psi)\, dD$$

by exchanging in the first term the subscripts m and n and also i and j and noting that $r_{ji} = r_{ij}$. In view of

$$e_i = - r_{ij}\, q_j \; = - r_{ij}\, \varepsilon_{jn}\, \partial_n \psi$$

this is also

$$- \int_D \varepsilon_{im} e_i\, \partial_m \delta\psi\, dD = - \int_S \varepsilon_{im} n_m\, e_i\, \delta\psi\, ds + \int_D \varepsilon_{im}\, \partial_m e_i\, \delta\psi\, dD \tag{5.1.13}$$

The second term in (5.1.13) will be set separately equal to zero, since it is the only one containing the arbitrary variation $\delta\psi$ within each finite element. The Euler equation is thus

$$\varepsilon_{im}\,\partial_m e_i = \partial_2 e_1 - \partial_1 e_2 = 0 \qquad (5.1.14)$$

It is the integrality condition for the temperature field; inside each element then this temperature field for which

$$e_i = \partial_i T \qquad (5.1.15)$$

will be determined except for an additive constant. Considering further that

$$\varepsilon_{im} n_m e_i = \varepsilon_{im} n_m \partial_i T = n_2\,\partial_1 T - n_1\,\partial_2 T = -\frac{\partial T}{\partial s}$$

the remaining contribution of (5.1.13) will be

$$\int_S \frac{\partial T}{\partial s}\,\delta\psi\,dS$$

and adding to it the variation of the second and third terms in (5.1.11)

$$\int_S \left(\frac{\partial T}{\partial s}\,\delta\psi + \frac{1}{h}\,\frac{\partial\psi}{\partial s}\,\delta\,\frac{\partial\psi}{\partial s}\right) ds + \int_{\partial_1 D} \bar{T}_e\,\delta\frac{\partial\psi}{\partial s}\,ds = 0$$

The integrations by parts on S are obtained by circulating anticlockwise around each finite element, which covers all the faces of S . For each element we can replace

$$\oint \frac{\partial T}{\partial s}\,\delta\psi\;ds = -\oint T\,\delta\frac{\partial\psi}{\partial s}\,ds$$

and obtain

$$(5.1.16) \qquad \int_S \left(\frac{1}{h} \frac{\partial \psi}{\partial s} - T \right) \delta \frac{\partial \psi}{\partial s} ds + \int_{\partial_1 D} \bar{T}_e \, \delta \frac{\partial \psi}{\partial s} ds = 0$$

On each interface

$$\left(\frac{\partial \psi}{\partial s} \right)_+ = - \left(\frac{\partial \psi}{\partial s} \right)_-$$

and the same holds for their variations, hence

$$(5.1.17) \qquad \left(-T + \frac{1}{h} \frac{\partial \psi}{\partial s} \right)_+ = \left(-T + \frac{1}{h} \frac{\partial \psi}{\partial s} \right)_- \quad \text{on interfaces}$$

which is equivalent to formula (2.7) of reference [1] . The variation of $\partial \psi / \partial s$ along $\partial_1 D$ yields

$$(5.1.18) \qquad T = \bar{T}_e \qquad \text{on } \partial_1 D$$

Finally the variation of $\partial \psi / \partial s$ on $\partial_2 D$ vanishes by virtue of (5.1.12).

The discretization of the scalar ψ follows similar rules to that of T in the temperature models, identification of nodal ψ values implying complete continuity of ψ across interfaces and consequently heat flow diffusivity. The possible $\bar{q}_e$ distributions along $\partial_2 D$ are of course limited by the shaping functions introduced to represent ψ . By contrast any distribution $\bar{T}_e$ along $\partial_1 D$ will be translated into generalized temperatures, linear functionals of $\bar{T}_e$ resulting from the discretization of the last integral in (5.1.11). An example of translation of boundary requirements in terms of the heat stream function is given in Figure 3.C.

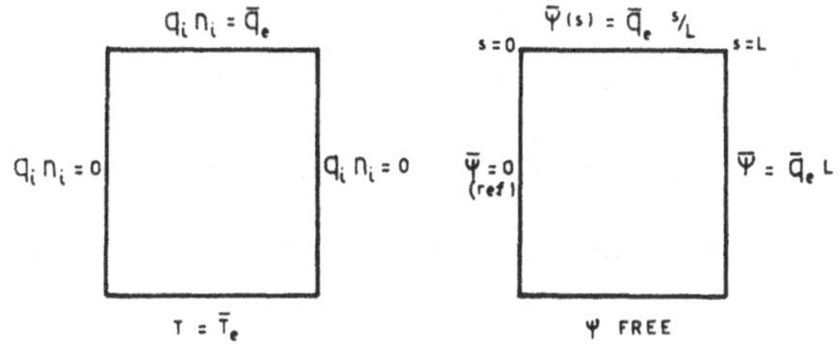

Fig. 3.c.

5.2 Three-dimensional heat flow models based on a vector stream function

The vector stream function a_i such that

$$q_j = \varepsilon_{jmn} \partial_m a_n \tag{5.2.1}$$

where ε_{jmn} is the three-dimensional alternating symbol, allows to satisfy "a priori" the homogeneous heat balance equation

$$\partial_j q_j = 0$$

(5.2.1) is equivalent to

$$\vec{q} = \text{rot } \vec{a}$$

and since rot grad $\Phi \equiv 0$

there is an indetermination in $\vec{a}$ constituted by the addition of the gradient of an arbitrary function.

It is easily seen that continuity of $\vec{a}$ is sufficient for heat flux diffusivity across interfaces. If a face portion is parallel to axes 2 and 3 , the heat flux density is $q_1 = \partial_2 a_3 - \partial_3 a_2$ and it depends only on derivatives in directions tangent to the face portion. More generally

$$n_j q_j = \varepsilon_{jmn} n_j \partial_m a_n$$

depends on derivatives in face tangent directions, since $\varepsilon_{jmn} n_j \partial_m$ is the vector product of the normal and the gradient operator.

Again in a domain without internal cavities $\vec{a}$ may be taken to be continuous throughout. In the presence of internal cavities with non zero penetrating heat flux, surface discontinuities must be organized and the method loses much of its interest.

Figure 4 presents some of the simplest heat flow models with their two types of discretization; it can be easily verified that the heat stream function approach leads to finite elements topologically identical to the corresponding temperature ones for 2-D approximations. The case of 3-D models is slightly different: their topological analogues are the class of the simplest 3-D conforming displacement models of elasticity.

2 - D Heat flow elements

$q_1 = \alpha_1$
$q_2 = \alpha_2$

$\phi = a + bx_1 + cx_2$

$q_1 = \alpha_1 + \alpha_2 x_1 + \alpha_3 x_2$
$q_2 = \alpha_4 + \alpha_5 x_1 - \alpha_2 x_2$

$\phi = a + bx_1 + cx_2 + dx_1^2$
$\quad + fx_1 x_2 + gx_2^2$

3 - D Heat flow element

$q_1 = \alpha_1$
$q_2 = \alpha_2$
$q_3 = \alpha_3$

$a_1 = \alpha_1 + \beta_1 x_1 + \gamma_1 x_2 + \delta_1 x_3$
$a_2 = \alpha_2 + \beta_2 x_1 + \gamma_2 x_2 + \delta_2 x_3$
$a_3 = \alpha_3 + \beta_3 x_1 + \gamma_3 x_2 + \delta_3 x_3$

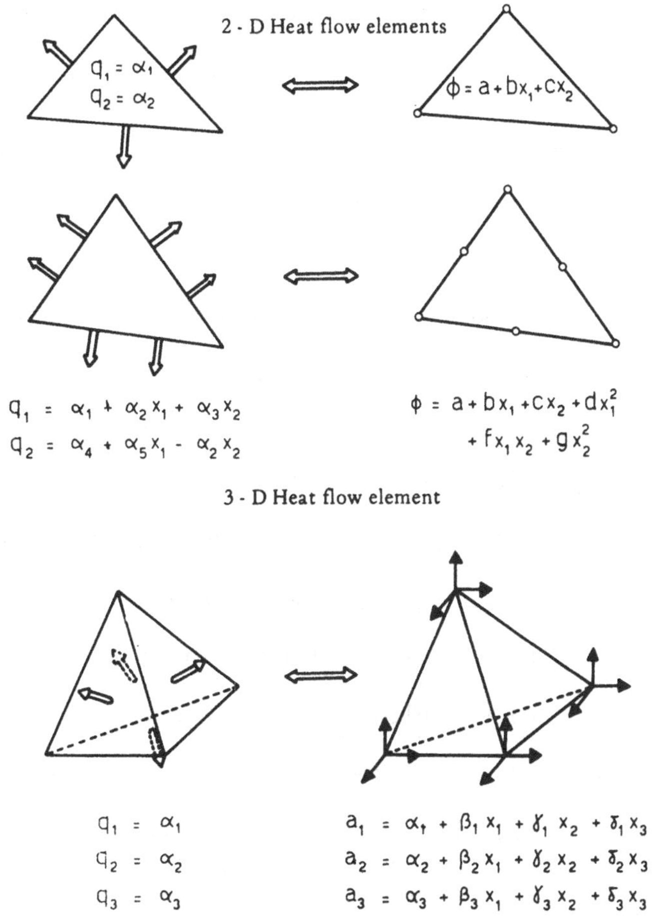

Fig. 4. Discretization of heat flow models

6. BOUNDS TO THE DISSIPATION FUNCTIONAL

The upper and lower bound character to the dissipation functional of approximations based on either the principle of variations of temperatures or the principle of variation of heat flow has been fully demonstrated in preceding developedments [1] , [2]. To obtain guaranteed bounds it is necessary to split the general problem in two parts, one where the prescribed temperature of the effective heat sink distribution is set equal to zero (problem 1) and one where all the prescribed heat sink rates vanish (problem 2). In the following two examples of type 2 will be briefly described. The awaited bounds for this category are an upper value for temperature models and the reverse for heat flow models.

6.1 Conduction-cooled turbine blade

A dual analysis of the blade described in section 4.4 (Fig.2) has been performed and yields correct upper and lower bounds. The mesh of figure 2 was used for each case of analysis and the different models together with their associated dissipation value are listed in the table hereafter.

FINITE ELEMENT MODEL	NB. DEGREES OF FREEDOM	10^{-4} D
Linear temperature field	23	20.91
Parabolic temperature field	67	20.53
Linear heat flow field	88	20.47
Constant heat flow field	44	19.55

6.2 Laminated slab (Fig. 5)

A laminated slab with orthotropic conductivities k_1 and k_2 is analyzed. The lateral surface exchanges heat with an isothermal fluid bath at T_f except the lower edge which is partly adiabatic and partly in contact with an external heat source at T_0. The detailed data are

$k_1 = .8$	$h_1 = .9$	$T_0 = 1{,}000$	$a = 10$
w/cm°c;	w/cm²°c;	°c;	cm.
$k_2 = .4$	$h_2 = .5$	$T_f = 100$	$b = 6$

Figure 5 presents the temperature maps for several models of approximations and the associated bounds are summarized in Figure 6. The cubic temperature model-treatment is nearly exact with respect to bounds bracketing. Heat flow models tend to converge faster than temperature ones but they need more degrees of freedom for a given grid an equal degrees in the

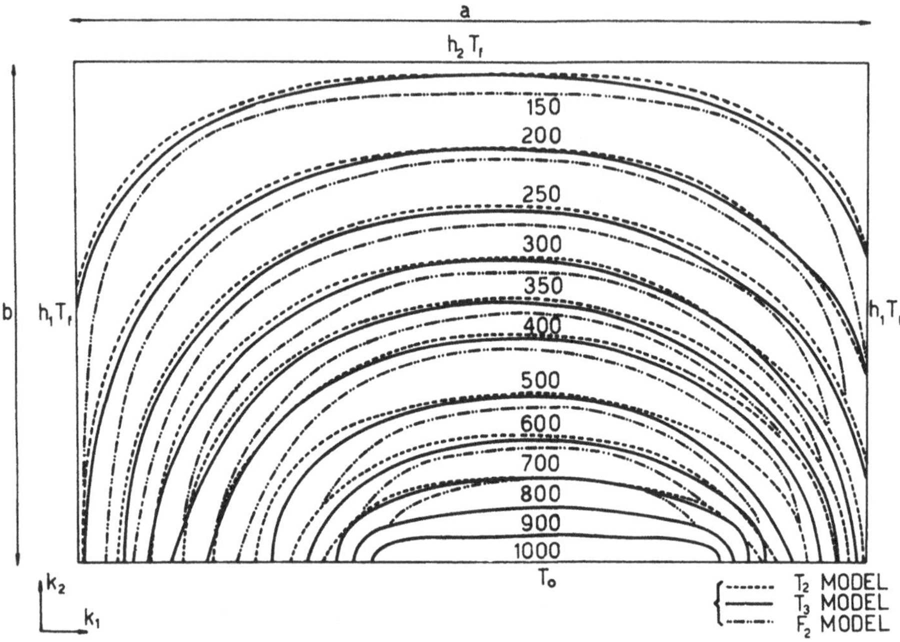

Fig. 5. Temperature distributions for the slab

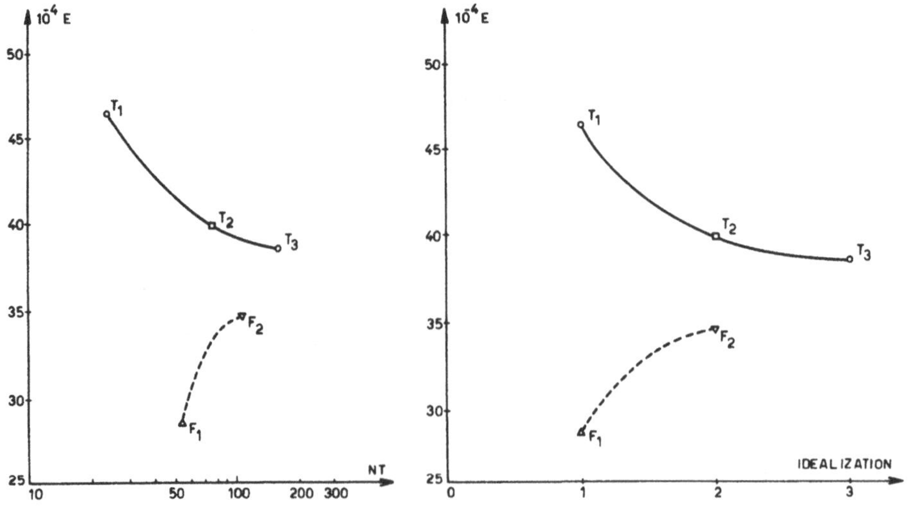

Fig. 6. Bounds to the dissipation functional

polynomial discretization of the element since they involve interface unknowns solely.

REFERENCES

[1] B.M. Fraeijs de Veubeke and M.A. Hogge, "Dual minimum theorems in heat conduction problems and finite element solution based on compatible temperature fields", CISM Course on the Finite Element Method, Udine, July 1972.

[2] B.M. Fraeijs de Veubeke and M.A. Hogge, "Dual analysis for heat conduction problems by finite elements", Int. Jnl for Num. Meth. in Eng., Vol. 4, 1972.

[3] M.A. Hogge, "Analyse numérique des problemes thermiques en construction aéronautique et spatiale", Report SF-14, L.T.A.S., Université de Liège, 1970.

CONTENTS

	Page
Structural dynamics	1
Preface	3
1. Introduction	5
2. The variational principles of elastodynamics	7
2.1 Hamilton's principle	7
2.2 The canonical variational principle	11
2.3 The complementary energy principle of elastodynamics	17
2.4 Matrix formulation of the variational principle	22
3. Eigenvalue analysis	24
3.1 The kinematical approach	25
3.3.1 The Rayleigh quotient	25
3.1.2 Independent or maximum-minimum characterization of eigenvalues	27
3.1.3 Recursive characterization of eigenvalues...	31
3.1.4 Minimum of the Rayleigh quotient under constraints.................................	31
3.2 The equilibrium approach	33
3.2.1 Self-stressing and vibration modes	33
3.2.2 Properties of Rayleigh quotient.............	36
4. Finite element models	40

Page

4.1 The displacement approach 40

4.2 Structural stiffness and mass matrices 48

4.3 Dependent boundary displacements. Super-elements 49

4.4 The equilibrium approach. Element flexibility and inverse-mass matrices 52

4.5 Solution of equilibrium approach in terms of unknown displacements 56

4.6 Kinematical freedoms of equilibrium elements.. 59

4.7 Structural assembling of equilibrium models... 61

5. Eigenvalue analysis in the presence of kinematical modes 64

5.1 Introduction 64

5.2 Kinematical modes and deformation modes 65

5.3 Static equilibrium conditions 68

5.4 The projection operator A. Pseudo-inversion of K. 69

5.5 Unicity of a pseudo-inverse. Isostaticity constraints. 73

5.6 Iteration in semi-definite eigenvalue problems 78

5.7 Numerical computation of kinematical modes.... 80

5.8 Obtention of a symmetric iteration matrix..... 85

6. The reduction methods 87

	Page
6.1 Introduction	87
6.2 Static condensation of variables	89
6.3 Bound algorithms	97
6.4 The substructure technique	104
6.5 The coupling methods	112
7. Numerical applications of the dual analysis to plate like structures	118
7.1 Cantilever square plate	120
7.2 Point supported plates	123
7.3 Cantilever skew plates	125
References	132
8. Transient response methods based on modal analysis analysis	137
8.1 Introduction	137
8.2 Modal analysis of linear elastic systems	138
References	164
Appendix I	165
Appendix II	167
Appendix III	169
Heat conduction	171
Preface	173
Part I. Lower bounds to a dissipation functional by temperature compatible finite elements......	175

Page

1. Introduction 175

2. Field equations and boundary conditions 177

3. Virtual power and heat sink potential 181

4. Scalar product and dissipation functional 183

5. The dual variational principles 185

5.1 The principle of variation of temperatures 189

5.2 The principle of variation of heat flows 191

6. Bounding of the dissipation functional 192

6.1 Bounding of the dissipation functional in problem 1 193

6.2 Bounding of the dissipation functional in problem 2 195

7. Mathematical models of temperature finite elements 197

7.1 Temperature elements generation 198

7.2 Local boundary temperatures and shaping functions 199

7.3 Local internal temperatures and bubble functions 201

7.4 Generalized heat fluxes 202

7.5 Element conductivity matrix 204

7.6 Specific convection boundary elements 206

7.7 Thermal properties of the assembled elements... 207

8. Numerical examples 208

Page

8.1 Human body thermal model 209

8.2 Pressure vessel temperature distribution...... 211

References 214

Part II. Upper bounds to a dissipation functional by balanced heat flux models 215

1. Introduction 215

2. Summary of the basic principle 216

3. Mathematical models of heat flow finite elements 218

3.1 Heat flow elements generation 218

3.2 Generalized heat fluxes 219

3.3 Generalized temperatures 221

3.4 Generalized conductivity relations 223

3.5 Remarks on heat flow finite elements 227

4. Examples and numerical effectiveness of heat flow elements 230

4.1 Constant heat flow model 230

4.2 Linear heat flow model 231

4.3 Heat flow model with internal heat sinks.... 231

4.4 Numerical examples 234

5. Construction of heat flow models by means of heat stream functions 235

5.1 Two-dimensional heat flow models based on a stream function 236

Page

5.2 Three-dimensional heat flow models based on a vector stream function 243

6. Bounds to the dissipation functional 246

6.1 Conduction-cooled turbine blade 246

6.2 Laminated slab 247

References 250

Contents 251

Printed in the United States
By Bookmasters